Competitive Strategy
Analysis for Agricultural
Marketing Cooperatives

Competitive Strategy Analysis for Agricultural Marketing Cooperatives

EDITED BY

Ronald W. Cotterill

Routledge
Taylor & Francis Group

LONDON AND NEW YORK

First published 1994 by Westview Press, Inc.

Published 2018 by Routledge
52 Vanderbilt Avenue, New York, NY 10017
2 Park Square, Milton Park, Abingdon, Oxon OX14 4RN

Routledge is an imprint of the Taylor & Francis Group, an informa business

A CIP catalog record for this book is available from the Library of Congress.

ISBN 13: 978-0-367-00911-3 (hbk)
ISBN 13: 978-0-367-15898-9 (pbk)

Contents

PART THREE
Strategic Alliances: Cooperative Marketing Agencies
in Common

Tables and Figures

Tables

Preface

Agricultural marketing cooperatives face strategic choices that can make or break them as business organizations. Successful choices are well known. Examples include Ocean Spray's success with blended cranberry juices and the creation of Sun-Diamond, a strategic alliance among the California fruit and nut cooperatives to jointly distribute and promote their branded products in the nation's grocery stores. Failures such as the Dairymen's League, Inc., exit from milk processing and the demise of their once strong regional brand, Dairylea, have received less press attention but are well known in cooperative circles. In addition to dramatic successes and failures, there is a middle ground. In many instances businesses dodge or defer strategic choices. They muddle through opportunities and respond defensively to threats.

Competitive strategy analysis seeks to identify strategic moves that create sustainable competitive advantage. Competitive advantage has two general sources: increased cost efficiency or market power that comes from underlying structural features, including substantial market share, and product differentiation. Cooperative marketing practice over the past sixty years has reinforced the two original theories or rationales for farmers to organize beyond the farm gate to improve their incomes. The competitive yardstick theory, first presented by Edward G. Nourse, counsels that farmers should integrate forward into the marketing system to share in the profits earned there. Such integrated cooperatives serve as a yardstick for farmers. When markets are not competitive these cooperatives can pay higher prices to patron members. Other farmers will shift patronage from investor-owned firms, ultimately forcing such firms also to pay higher procurement prices. Nourse's yardstick theory was developed for undifferentiated commodity products. In modern markets where food products are differentiated and brand-name capital is important, the decision to integrate is more complex. Should the cooperative produce only private-label products, should it develop branded products, or should it do both? The second general rationale for cooperative marketing is Sapiro's horizontal integration of all producers into a commodity marketing association such as the raisin or cranberry growers. Control of all or nearly all of the commodity allows such an organization to capture real marketing efficiencies in assembly and possibly basic commodity processing. It may also give the cooperative market power vis-à-vis downstream buyers of the products.

In practice some of the most successful agricultural marketing cooperatives combine horizontal and vertical integration with the development of branded products. Other cooperatives have focused upon branded products that serve niche markets. Nonetheless, most agricultural marketing cooperatives sell predominantly unbranded products. In commodity areas such as milk, these cooperatives often control a very large share of the market.

This book introduces the subject of competitive strategy analysis for agricultural marketing cooperatives. It explores issues of horizontal and vertical integration and product differentiation by describing current cooperative practices and discussing new strategic directions that cooperatives might pursue. It is organized into three parts.

Part One discusses strategic planning from a business perspective. The first chapter is by a cooperative chief executive officer and the second describes the comprehensive restructuring of a large multiproduct marketing cooperative.

Part Two contains four chapters that address the underlying theory of vertical integration by cooperatives, the extent of product differentiation by agricultural marketing cooperatives, and a possible source of new differentiation unique to cooperatives—the ability to control product quality in the vertical production and marketing system.

Part Three explores a type of strategic alliance unique to cooperatives: the marketing agency in common. The four chapters in this section provide a general overview and describe three different particular examples: the dairy marketing initiative in the Upper Midwest, the Sun-Diamond joint sales effort by five California fruit and nut cooperatives, and the information-sharing cooperative in the California lettuce industry.

Ronald W. Cotterill

Acknowledgments

This book is based upon a workshop titled "New Strategic Directions for Agricultural Marketing Cooperatives," which was held in Boston on June 24-25, 1992. The workshop was jointly sponsored by Regional Research Project NE-165: Private Strategies, Public Policies, and Food System Performance; Regional Coordinating Committee NCR-140 on Cooperatives; the Agricultural Cooperative Service, USDA; the National Council of Farmer Cooperatives; the Cooperative Business Association; and the University of Connecticut Food Marketing Policy Center. A special acknowledgment is due to the workshop organizing committee that set the workshop agenda and selected papers from over 50 submissions. Members included Bruce Anderson, Cornell University; Julie Caswell, University of Massachusetts-Amherst; Edward Jesse, University of Wisconsin-Madison; and Randall Torgerson, Agricultural Cooperative Service, USDA. Special thanks to Andrew Franklin and Irene Dionne, Food Marketing Policy Center staff members, for their assistance in organizing the workshop and producing this book.

R.W.C.

## PART ONE

Strategic Planning:
A Business Perspective

1

Strategic Planning in Agricultural Cooperatives: Riceland Foods, Inc.

Richard E. Bell

I appreciate the opportunity to speak to you today about strategic planning. I have to admit in our company I know who is the Director of Strategic Planning. It's me! I believe that if the C.E.O. doesn't have a background in strategic planning, it's very hard to have an objective program. At the outset, I would like to discuss the need for a well defined strategic mission. If you don't know what your mission is, you can do little strategic planning. Strategic planning includes developing a plan that allows you to target your best shot. The secret of successful planning is identifying what your best shot is, and you must know what your mission is to do so. One of the key components of mission for a cooperative is that it is likely doing something that no one else wants to do. If its task was attractive to an investor-owned firm, ADM or Cargill, or another investor-owned firm would be doing it.

In fact, cooperatives are in grain and/or rice marketing because no one else wants to do it. In our market driven economy, others don't think it will pay them enough. So that puts a grain marketing cooperative in a hole right off in terms of mission.

I want to tell you about Riceland because it lets you know where I'm coming from. I have to admit as I was sitting in Stuttgart the last couple of days thinking about coming here, I was really startled by how Riceland has changed since I came on board in 1977. The first day I worked at Riceland I asked to see a copy of the budget. There was no budget. Next, I asked for the organizational chart. There was no organizational chart. I couldn't understand how anyone could run an organization like that, but run it did and it was doing well.

Riceland, first of all, is a marketing cooperative, and that is all we want to be. We want to market rice and soybeans, stay in our business lines, and stay in our geographic area. We don't want to be the biggest in the world. We don't want to be in California rice or Minnesota soybeans. We want to execute the mission which our farmers have paid us to do.

Riceland is a membership cooperative which is different from a federated cooperative. I have no question who I work for. I work for Riceland's 11,000 farmer members. I don't work for some other co-op manager or another cooperative. As a membership cooperative I believe Riceland is much more responsive to farmer-member needs and wishes than one of the other types of cooperatives. I'll come back to this as part of my strategy.

As a pool marketing organization, members' soybeans and rice go into pools which are marketed by the professional staff. Farmers do not make their own decisions about pricing. Their job is to produce crops and do the best job they can on the farm. It is Riceland's job to be their marketing agency. Our entire system is driven by two other factors, one of which is to market grains, as much as possible, in value added form. We have been selling wheat to the Chinese for ten years or so but don't like the discounts the Chinese demand for sprout damaged wheat. Riceland must garner some muscle in terms of being able to control channels of supply and price. We already have a rice flour mill that provides us with some supply channel control in domestic markets. Our next move will be to build a small wheat flour mill.

Riceland is integrated all the way from the farmer to the shelf in Chicago. We are not in the retail rice business here in Boston. We do, however, have a good food service business here. I was pleased last night when I was asked at a restaurant if I wanted rice or potatoes. The choice was easy!

To be successful in strategic planning one has to have a complete knowledge of the business. That's why I have trouble hiring somebody from the outside to come in and give advice about strategic planning. I don't think they are going to know as much as we know about our business.

Also, one must do hard-headed analysis which requires putting complex extensive data into a manageable form. I deal with grain traders every day. They have no idea how to do hard-headed analysis. All they want to do is sell, sell, sell. As CEO I must get them into the planning process, but I certainly cannot let them decide the strategic plan. They will tell me we should just sell grain. A cooperative such as Riceland must have a more far sighted mission that is based upon insight into the grain marketing business and the cooperative as a business organization.

Returning now to the fact that Riceland is a membership cooperative, it is important to realize that our membership, and I'm sure this is also true of the other cooperatives, is made up of individuals and corporations. The distinction is important. Concerning individual members, one of my favorite groups is the widows of members. If you think they don't own the co-op, you are wrong. They are an important stabilizing element in the cooperative.

Corporate members are also important. A corporation never dies. It may go out of business, but it doesn't die when the business is transferred from one generation to the other. Once a corporation is vested in the Riceland Foods base capital plan, there are no transfers. The corporate farm doesn't have to come back and keep investing when the farm changes hands.

Membership structure also is important for strategic planning because it determines the governance structure. We at Riceland have direct voting. Although everybody votes, we have a weighted voting system. The larger, incorporated farms in the plantation area that used to grow cotton and now grow rice have much more say than, perhaps, some of the widows. Weighted voting is based upon participation in the cooperative.

The other governance issue that is important for planning is the separation of the board and management. I have to admit that I have seen this problem in the U.S. Department of Agriculture as well as in cooperatives. When I was in the USDA, all the people in the South Building, the career people, knew how to make policy, and all those over in the Administration Building knew how to do operations. As we used to say, Independence Avenue ran in between, but it was mostly just a "mishmash". It's important that these functions be separate. Making policy sounds like fun, but it is difficult, just as strategic planning is difficult. Managers who are in charge of operations should not get into policy making.

At Riceland we have been very blessed. Our board hires management to run the operation. They set policy and we implement it. I have to admit that they come back from time to time and say, "Tell us again what our role is? What do you mean by setting policy?" We tell them that basically it is making long-term financial decisions, concerning revolving equities, hiring management and setting broad policies. It isn't deciding what kind of pickup we'll buy. That is a management decision.

Some observers claim that I have a strategy for everything. That may be true. Let's talk about financial strategy, marketing strategy, origination strategy, and processing strategy. They flow from our mission to market the rice and soybean products for our farmer members and to return the best price that we can. It's as simple as that. Given the basic mission, our strategy determines how effective we are.

Financial Strategy

I'm a great believer in having a strong equity base. In our industry a strong equity base is 85 percent equity to total permanent assets. Why do we need to have an asset ratio like that? That's easy. Look at ADM, Cargill, and ConAgra? They are all 85-100 percent. You must have that type of ownership commitment to a grain marketing business or you have no chance of having a successful strategy. It took us almost 10 years to get to 85 percent, because we started at 40.

Whitney McMillan, the Chairman of Cargill recently said that in order to be successful, your company must double in size every five to seven years. What he's talking about, however, is not sales. He is talking about net worth—the value of the business to its shareholder-owners. It is the most telling measure in terms of whether you're being successful or not. After reading his statement, I looked at Riceland's performance. We haven't doubled every 5 to 7 years, but we are not far from that standard.

Having a strong equity position provides debt flexibility. I want to have that 85 percent ratio. I have to admit that it doesn't have to really be that high. By having it, however, I have the option to go out and reach for something that I think is important. It always gives the opportunity for one more acquisition or one more project which really might help the cooperative.

I have to admit I'm a big supporter of the CoBank. Why? The rates are good! They are competitive with other lenders and if we were in the money market we'd be a triple "A" industrial which allows us to obtain very low rates. When we go into the money market to check the Cobank, with a little prodding, they're right where they're supposed to be.

Moving beyond debt flexibility to capitalization, our plan is called a base capital plan, which is familiar to those of you in the dairy business. Our members have a seven year period to make a capital contribution which is commensurate with their participation in the cooperative from their beginning benchmark number. It's rather simple: $1 is invested for every bushel of soybeans and rice that they deliver. Frankly, as of today, they're almost 100 percent vested. We have about $85 million in base capital. I view that as my common stock. It's going to be there unless we really foul up. Individuals may leave the business, but the incorporated farm businesses never leave the cooperative because a corporation never dies. The component of permanent capital, which is really quite different from when I first came, is retained earnings. Earlier we had a revolving stock program. Now we try to make money, pay the tax and put it into a permanent capital fund. Originally, our goal was to get our total retained earnings to be about 15 percent of total equity. When we got to 15 percent I said I'd like to get to 25 percent; and, now that we're up to 30 percent I'm ready to go to 35 percent. Someone asked me how high was I going to go? I'll probably go as high as the board will let me. Again, retained earnings give flexibility. Riceland has the capital to do for its 11,000 members what needs to be done. They do not have to come back and put up another dollar. In general, the membership views its $85 million base capital plan contributions as all that they want to contribute.

Dr. Charles Black at Texas A & M used to really rake me over the coals, because he said I'd go to 100 percent returned earnings if I could. I said, "No, maybe 85 percent but not 100 percent. Again, in order to be competitive with the Cargills and the ConAgras, we must have a strong equity position with substantial permanent capital from retained earnings. That's who Riceland competes with, not with another cooperative.

Marketing Strategy

Marketing strategy is important in any business, but I think it is critical in grain marketing. One of our keys to success in marketing is that we can provide Kellogg or General Foods with a continuous supply. We're in the market every day. Other suppliers have to buy grain from the trade. We have the raw material. It's delivered by those who grow it, so we have a continuous supply and should have stable prices throughout the year. Part of our marketing strategy also is diversification. I want to be in as many markets as possible every day. That provides a degree of stability in terms of the business. This year the rice market is not that good, but the soybean market is.

A firm must also have meaningful shares in the markets where it sells. If you don't have meaningful shares, you are just an also-ran, and you're not going to gain much attention. This is particularly important when competing at the retail level. If a company's branded product is not first or second, it probably should not be there because it is not going to get the retail distribution it needs to succeed. I think that is also true in food-service distribution channels, and it's true for the ingredient supply business which is becoming an important business line.

We supply four major market segments and have ten separate business lines. I talk to my staff about each one every month to monitor how we're doing in those business lines. The four market segments that the ten business lines are in are retail distribution, food-service, ingredients, and the export market.

Riceland has been a major international player for a long, long time. We have had some very unusual customers. At one time I loved Iranians and I was probably the last to give up on Saddam Hussein, because Iraq was one of our major markets for rice. Today, we're moving away from some customers that are politically unstable. Generally, our business has been about half international, but this year our rice business is going to be 70 percent domestic which tells you how the markets have changed.

With regard to retail distribution, we are not a national brand. Our Riceland Rice label is a regional brand. We're first in Chicago, but we're not in Boston. We're not here because there's no way that I can get into this market and be first or second. But if somebody wants to sell private label rice in Boston, or elsewhere, we'll pack it.

In the food-service segment, much of our business has been with distributors to Chinese restaurants. The nice thing about the Chinese is they're always with you; the bad thing is that they rarely pay promptly. The other part of food service that has developed and has become extremely important to be in, is the area that we call the national distributors of food-service products. These firms include SYSCO, Sexton, and Kraft. They have emerged to be the major players in this sector. If Riceland is not selling to them, it is not going to be in food-service. In fact, we got in trouble with SYSCO because we wanted to distribute our Chef-way Rice. They don't want to use Chef-way rice they want to use SYSCO rice. We quickly realized that we were going to move a lot more rice with a SYSCO label.

Another market segment where we've had really good growth is ingredients. When a consumer goes into a retail outlet and looks at all the rice products on the shelf, she may or may not buy the Riceland label but you can be sure that almost every one of the other rice-based products has our rice in it. This includes Ricaroni, General Foods Minute Rice and even Rivianna's Mahatma.

The other ingredient area that's really been important is providing rice to food manufacturers. For many years our number one market for rice was Anheuser-Busch. Every time those Budweiser numbers came out in the 1980s, I really glowed. The only thing that's happened since is that the dog food industry is replacing Anheuser-Busch as our top ingredient buyer of rice. I don't know if you've ever looked at dog food labels. Most of them now have rice. It makes the doggies tummies feel good. It's not allergenic. The dog food firms have more exacting standards, but they pay a better price than Anheuser-Busch.

Origination Strategy

Origination strategy is making sure that you get the raw material. Basically, our contracting system is voluntary. Unlike the California system of committed production, our members do not have long term contracts with Riceland; so I'm always promoting Riceland to our members. One of my activities in February will be going to membership meetings with farmers at 7:00 a.m. We have breakfast meetings. My favorite is on the Missouri/Arkansas line where we meet at the Boss Hog's Stateline Cafe. Not many people have been there. But I go out there and talk up our membership—then what I'm doing for them, and that I have the right price and the right service. In our view, service really only means one thing in the end—dumping trucks. Riceland has to dump those grain trucks and turn them around at harvest time. We also do their paper work. That's why the widows love us. If they get in trouble with the I.R.S., we're there to help them. We have the types of systems to do that. In summary, Riceland must give a premium return which means they do better with us than they can with somebody else. That's how we measure premium return.

Processing Strategy

One of Riceland's general strategies is to be a low-cost manufacturer. That requires modern manufacturing practices in today's world. We have really worked hard to make that happen. We already have installed order-driven manufacturing which is coupled with computer-integrated manufacturing and just-in-time inventory management. I told people for a long time we had the "just-in-case" system. We made it "just-in-case" somebody called. Now we manufacture to order and ship it out just in time to get it on the truck. It has had a big impact on our costs. We also have quality assurance programs that are based upon statistical process control. As a result, costs are cut and there is a tremendous increase in operating efficiency.

Riceland's management information system is based upon LANS (local area network) technology. Our system includes orderdriven manufacturing. We are at the point where we cannot run without the computer. We have a disaster recovery contract. If we have some type of natural disaster and the computer can't run, we declare a disaster and can be up and running at another location in a 24 hour period.

Another feature of the LANS that I want to mention is that it is connected with a wide-area network. At each of our local offices all our member information can be called up on the screen from the mainframe computer to display individual member investments and account activity.

Shifting now to more general issues in the strategic environment, I want to talk about economic policy. When I go around the country I don't get the impression that we're going down the tubes as some people contend. I think that we're doing a lot of things right. Look at our export numbers, for example. I think that we have considerable economic strength. We have great technological advantages. We're leading in most of the technological areas that really count and we have the flexibility to change which, again, is very important in terms of international competition.

Yet, in agriculture, we have policy constraints that I briefly want to talk about. One of those is the budget. We've peaked out in the amount of money that we're going to spend on agricultural programs. So, when formulating a strategy, don't rely heavily upon help from the government. Agriculture and agribusiness are going to have to do it on their own, which I, in fact, like. We still have the problems of divergent regional interests: the South, the West Coast (which is California), the Mid-West, and the Northeast. We must do a better job of communicating what we're doing.

As far as the GATT is concerned, I think that any possibility for progress on free trade ended a long time ago. We have just not accepted it. I'm a big believer in the North America Free Trade Area Agreement. I believe a lot of our growth is going to come in the north-south, not east-west trade. I think my opportunities are greater in Mexico than in the former Soviet Union or in Eastern Europe.

On the emerging issues, I have great concern in terms of our agricultural policy and our leadership. We, in fact, are too domestically oriented at the present time. I say this even though our business is 70 percent domestic. Our recent shift towards domestic sales is due to a rapid increase in ethnic groups that have come into the country. We have had tremendous growth in our rice business in the hispanic areas: South Florida, South Texas, and South California. We want to get Cuba back as a major rice market, but if we don't get it back soon, the only guy left there will be Castro.

Nonetheless, I think that from an agricultural standpoint, international markets are important to us. We must be competitive and get out there and search for new markets. That has become an important part of our future.

On that note, I will say a few words about value-added exports. Everyone talks about value added exports. There are several different definitions of value added. To me, it's anything we do that gives Riceland some edge over the competition. It doesn't have to be a branded retail product to be value added. The ingredients that we're selling to the dog food companies are value added. They won't buy those ingredients from just anybody. Anheuser is going to buy rice ingredients from Riceland because we have quality assurance and dependability. We charge them for it, and that is value added.

You can do the same thing in exports. We have a nice business in Saudi Arabia. I always was glad that Saddam Hussein stopped where he did. We have a value added export rice business in Saudi Arabia where we're selling rice in a package that's special to Saudi Arabia. Another firm has a "one girl" rice. We have two Arab girls on our package, so our product is known as the "two-girl" rice. Many Arabs can't read, but they can tell two girls from one girl.

Business Alliances

Again, I really am amazed how business alliances have begun to emerge in recent years as an important part of strategy. When we look at forming an alliance, the partner is rarely a cooperative. It's somebody who is in the same business that brings an element of strength that we don't have. Joining in a business alliance can really make things happen. We already have a couple partnerships where we're doing manufacturing or marketing. We have one with AGP for lecithin. Lecithin is a by-product from soybean crushing. There are only three manufacturers of lecithin in the world. There is one in Germany and two in the United States, and we're one of them. We're marketing the product worldwide. The raw material we're using is coming from AGP which is the co-op that processes soybeans in the Midwest. I see business alliances as growing, growing, and growing because it's almost a Japanese system—almost, but not quite.

Direct Contracting

Direct contracting is related to commodity exchanges. I'm probably the only member of the Chicago Board of Trade here. I think that the usefulness of the commodity exchange, from an agricultural standpoint, peaked some time ago. Did you notice that the other day they started GLOBEX, a worldwide electronic trading system. What contracts do they want to trade? They offer only financials. They're not going to trade agricultural commodities. One of the problems I think that we have in the commodity exchanges is that we in agriculture have become such a small part of them. We've lost control. The real power lies with those who trade in the financials and other nonagricultural contracts.

There also is tremendous participation by mutual funds, pension funds, and other institutional investors. They have tremendous economic power, and they can move prices from day to day which causes instability for wheat and agriculture. So I stay away from exchanges as much as possible although we're a member of the Board of Trade and use it for hedging but not as much as at one time.

I keep stressing that one can do the same thing as a commodity exchange with marketing pools. Pooling spreads risk and aids price discovery. It does all the things that farmers and agribusiness are doing when they use the exchanges in Chicago, but, in this case, farmers and their cooperatives are in charge.

Direct contracting at Riceland involves having our farmers sign up each year so we know whether they are going to commit to us. Once we know that, we know the size of that pool. We know what the risks are and that serves all those same functions you have in a commodity exchange. A direct contract pooling system may be an attractive institutional alternative to commodity exchanges.

That's my presentation. If anybody has any questions or comments I'd be happy to at least try to answer them.

Questions and Answers

1. What percentage of the soybeans would you buy from your members? In other words, if the members produce "X" bushels, what percentage do you buy? About 40 percent of our beans go into what we call pools. We don't buy that grain, we just act as the marketing agent. The other 60 percent of our business is on buy/sell contracts. Frankly, that 60 percent is as high as it is because we have terminal elevators on the Arkansas River and the Mississippi River that are not in our pool areas. These are old cotton areas. But in the established rice growing areas where growers are accustomed to the seasonal pool marketing, soybeans are almost 100 percent in pools.

2. Is it 100 percent of their production or 100 percent of what they give to

sell to you that goes into the pool? Generally, if you are a pool person you are 100 percent. As individuals there are very few who are both buy/sell and seasonal pool, but a few.

3. How many pools do you run? It depends on the accountants. We have ineligible and eligible pools for price support purposes. We have them in rice, soybeans, wheat, corn, milo, and oats. But within each commodity we'll have several pools. One of the soybean pools is for growers who make a preharvest commitment. This gives Riceland an opportunity to do more hedging before harvest. Other pools allow growers to wait and make their decision after they have delivered their crop.

4. Would you elaborate on Riceland's shift to weighted voting? When I first came to Riceland, we didn't have weighted voting. After I went through one annual meeting where we had people complaining who had a very small stake in the co-op, I knew we had to have something else. It was possible to come to one of our annual meetings as a membership cooperative with a few people who delivered a few bushels and actually decide things. So, we decided we had to have weighted voting related to the person's investment and their deliveries to the cooperative. We put that system into effect about a decade ago and it has worked fine.

5. Do you do weighted voting for board members? When we elect a board member it is done by weighted voting. But it is done by district and districts are established according to weights. But if you're doing your business right, you're not going to have many contested elections. As a cooperative moves along, it ought to spot leadership early on and have a leadership training program to develop that leadership.

6. You said that the GATT is not going to produce much of anything for agriculture. Why do you believe that? Is there anything there or should we just walk away from it? Well, I don't think that the politics were correct at the start for meaningful negotiations. That's strategic targeting again. The timing of it, in my view, just wasn't there because of what was happening in Europe. Europeans were busy building Europe but not busy building the world. To get them to focus on international negotiations at this time, I think is impossible.

There are some things, however, in sectors other than agriculture, that I think are important. One is intellectual property rights. We've had a real problem in our rice marketing. There is a firm in Thailand that calls itself Riceland International which is an infringement on our name and products. A man who attended the University of Michigan went home and then started Riceland International. There's no way to stop him because we don't have an international treaty on intellectual property rights. If you go down to Panama, you're probably going to see Riceland Rice there but on the bag it will say "Riceland International." That irritates me. It's even the same color, almost identical. It showed up in Sweeden but the Swedes took care of it. An international intellectual property rights agreement, I think, would be important for American business.

7. You mentioned the idea of pools as an alternative to the futures market. I had to smile when you said that because probably the last major academic writing on grain pools was done by Joe Knapp for the hard wheat winter pools in the 1920s and 1930s. More recent work by McCalla and Schmitz in the 1970s compares the Canadian grain marketing system; i.e., Canadian wheat board pools, to future exchange based marketing in the U.S. I was wondering if you would comment a little more on that or do you just want your members to not hedge in order that they will join Riceland's pools? I wonder if you have a bigger vision of the issue? Well, the Canadian wheat board system is no different than our rice marketing system and, of course, now it's under pressure as is the one in Australia. That doesn't mean that we're not going to have Saskatchewan wheat pool or the Manitoba wheat pool or the Alberta wheat pool. As you suggest, their involvement in futures trading is, in my judgement, very small. I don't think the Australian wheat board or the Canadian wheat board have been major players in futures trading. I feel that the futures market system grew up at the time it did because something had to be done to bridge the middle. I don't think that necessarily exists today. Why do you still have trading pits in Chicago? Today, there is computer trading. People in Chicago don't want to give up the pits because it is to their advantage, but they will have to soon. We can do computer trading without the exchange pits. I just see them fading away in time but they will die slowly.

8. Could you speak a little bit about the prospect of doing business with the Japanese; i.e., selling rice there? We have two Japanese projects and one of those, in fact, is with a firm in Japan to sell rice mixes. You can sell rice there now as long as you mix it with something else. We have products we're developing which we hope eventually can get into the Japanese market. As far as the issue of rice between the United States and Japan is concerned, it's more of a West Coast issue in my judgement than a South issue. West Coast growers have been thinking that they, in fact, would be able to sell their brand grain rice or medium grain rice in Japan as a replacement with Japanese grown rice. Frankly, again, if we ever get such broad access that's not going to be my strategy at all. My strategy would be to get them to buy high quality Southern long grain rice which would be in blends. We're not going to replace their rice. But they have Kentucky Fried Chicken and McDonalds. They'll pick up a lot of our worker ways of doing things, particularly the younger generation. So I could see us selling our Riceland long grain rice products in Japan. My strategy is to develop a partnership and gain entry for a smaller segment of the market. I don't know when we're going to get there because, again, that was going to be one of the accomplishments that would come out of the GATT negotiation. Without GATT I don't know if we'll really be able to enter on a bilateral basis.

9. Do you have an estimate of how much rice would be grown in the United States without price support programs? How does that fit into your long range strategic planning? There would be less rice grown in the United States without the price support program, but most of it would be grown along the Mississippi River which, again, our strategy takes into account. If you look at the cost of production of the various regions that grow rice, Texas is already receding as a major rice growing area. One of the things I don't like is what we call 50/92 in terms of the price support program. This is where you only grow 50 percent of your rice but you get 92 percent of your price support benefits. I don't think the government should pay somebody for something they weren't going to do anyway. I think that's happening in Texas. They can't grow rice without the 50/92 payments. The base is declining there and now it is also beginning to decline in Louisiana. But, as far as the eastern part of Arkansas, northwest Mississippi, northeast Louisiana, southeast Missouri, I think we could be competitive on a world basis. I think that it would be a smaller industry but it would be in the area that I talk about. Also, we're diversified, rice isn't our only option. You notice that I haven't mentioned California because I think California has been having a continuing struggle in terms of maintaining a viable rice industry at the size that they've known, both for water and environmental reasons. Thanks.

Notes

Mr. Bell appeared as a guest speaker at the NE-165 Workshop *New Strategic Directions for Agricultural Marketing Cooperatives*, Boston, Massachusetts, June 25, 1992. This paper is transcribed from that speech.

2

Reconfiguration of the Coopérative Fédérée de Québec for New Strategic Directions

Randall E. Westgren

In 1989, the Coopérative Fédérée de Québec completed an extensive strategic planning process which included the input of its member coops and individual members. The mission statement and subsequent decisions based upon this planning process reflect a significant change in how the Coop Fédérée envisages its strategic position. During the 1990 and 1991 fiscal years, the company formed joint ventures in dairy processing and meat packing, purchased the shares of Tyson Canada, and purchased $15 million (Can) of assets in vegetable marketing and farm supply.

These strategic decisions result from a variety of threats and opportunities engendered in the operating environment of the divisions of the Coop Fédérée. The confluence of the Canada-U.S. Trade Agreement (CUSTA), the imminent conclusion of both the North American Trade Agreement and the GATT negotiations, Canadian agricultural policy and market regulation that limit strategic flexibility of processing firms, and the rupture of the relationship between dairy farmers and their cooperatives has forced the Coopérative Fédérée to reconfigure its structure.

This paper is written as the company is still wrestling with the problems of assimilating new investments, rationalizing wholly-owned and subsidiary organizations, and learning how to participate in two large joint ventures. One of the difficulties with researching the relationship between firm performance and strategic decisions is that the observability of cause and effect is confounded by dynamics and by the multiplicity of concurrent decisions. In this case, the Coop Fédérée has simultaneously made several strategic decisions that will take two or more years to fully implement, by which time their strategic outcomes will be affected by a myriad of operational decisions and changes in the operating environments of the various strategic business units. Thus, it will take some effort to measure the performance consequences of this reconfiguration.

The paper proceeds in three sections. First, a brief history and description of the Coopérative Fédérée de Québec is presented. This is followed by a discussion of the remote and task environments of the strategic business units in processing and marketing of agricultural products. Inferences for strategic choice will be drawn from this discussion. Finally, the strategic choices taken in the 1990-1991 period will be analyzed in the context of the mission statement of the company and the discussion of the operating environments.

The Coopérative Fédérée de Québec[1]

The Société Coopérative Fédérée des Agriculteurs de Québec was founded in 1922 by joining a seed marketing coop (La Coopérative des Producteurs de Semences de Québec), an affiliation of dairy processing coops (La Coopérative des Fromagers de Québec), and a farm supply coop (Le Comptoir Coopératif). These three cooperatives were founded in the period 1910-1914 and had a total membership of 6,800 members from among the 145,000 Québec farmers. The federation was formed under special legislation which on one hand permitted it to be the sole federation of cooperatives in the province, but also gave the ministry of agriculture veto power over its decisions and the right to name the managing director. In 1929, the law was amended and these restrictions removed.

The farm supply operations of the cooperative grew constantly from the 1930s onward and in 1966 the farm supply group was established. It included feed milling, fabrication of fertilizers, distribution of chemicals and seeds, and farm advisory services for livestock and crop producers. The group also consisted of Co-op stores owned and operated by the Fédérée, and a wholesale machinery and hardware operation that began in 1950. Petroleum wholesale and retail operations began in 1958 and a separate operating division was organized in the mid 1980s.

The dairy division was also created in 1966 from the marketing services of the Fédérée on behalf of the constituent local cooperatives, its own processing plant, and various distribution facilities across the province. This reorganization accompanied the height of the consolidation of dairy cooperatives in the mid-1960s, which was facilitated by Bill 72, a government program to subsidize the closing of small plants and to increase the scale of others.

The Coopérative Fédérée continually built and purchased livestock slaughter facilities from the inception of the company. This activity has always been organized as a centralized activity, with the assets owned by the Fédérée. Pork became the dominant livestock activity for the abattoirs in 1975, as this subsector of the Québec agricultural economy started its growth towards 18%

nf all farm receipts in 1991, second in scope only to dairy (Service de l'information et des statistiques, 1992).

The Poultry Division began as a marketing service for live and farm-slaughtered chickens in the 1920s. In 1975, the Coop Fédérée purchased Québec Poultry, Ltd., which was renamed Bexel. This division consisted of growing facilities, hatcheries, primary processing, and secondary processing.

The Fédérée has had fruit and vegetable marketing operations on a small scale since the 1940s. Over time, the local coops in the horticultural regions (usually mixed marketing-supply coops) took over the marketing and export activities that had been handled by the federation. The value of assets in this area was minuscule by the late 1970s.

Company Structure in 1989

At the time of completion of the strategic planning exercise in 1989, the company had finished five years of increasing revenues to a record level of $1.27 billion in 1988 (Figure 2.1), had three years of historically high levels of operating savings in 1986-1988 (Figure 2.2), had four years of increasing patronage refunds to an all-time high of $12 million (Figure 2.3), and had six

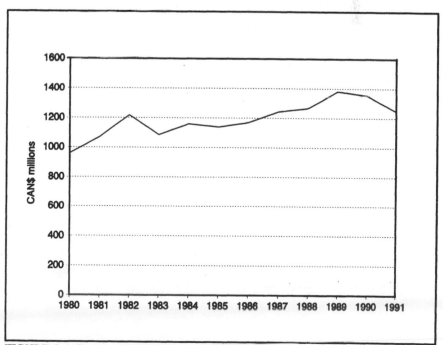

FIGURE 2.1 Coopérative Fédérée Total Revenues

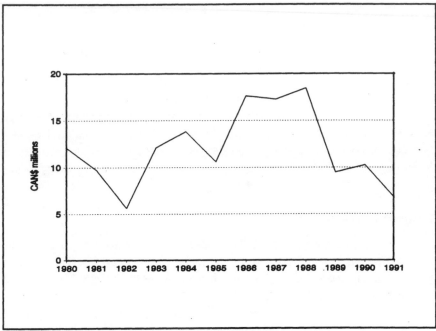

FIGURE 2.2 Coopérative Fédérée Operating Savings

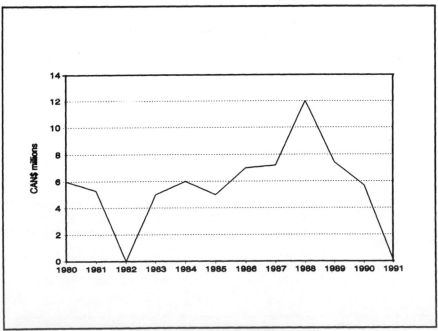

FIGURE 2.3 Coopérative Fédérée Patronage Refunds

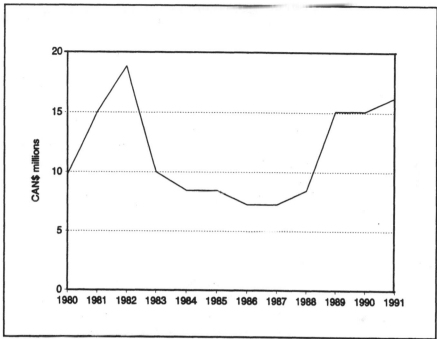

FIGURE 2.4 Coopérative Fédérée Financing Costs

years of low financing costs during the period 1983-1988 (Figure 2.4), The Fédérée was well positioned financially to make a change in strategic direction.

At the beginning of 1989, the Fédérée operated as a mixed federated-centralized cooperative. The ownership of the Fédérée was vested with 99 local cooperatives primarily in the supply sector, 8 regional milk cooperatives, and 2221 direct farmer-members. This vast majority of this latter group was affiliated with two marketing divisions that acted as centralized regional cooperatives: the meat division (Turcotte et Turmel) and the poultry division (Bexel). The other operating divisions were Petroleum, Dairy, the Farm Supply Group, and the Fruits and Vegetables Department. The Farm Supply Group consisted of feed milling and sales, grain sales, fertilizer production and sales, animal health products, seeds, agricultural chemicals, hardware and farm machinery, and seven retail outlets. This group had total revenues of $483 million in 1988 and $518 million in 1989, which represented 30.2% of total revenues of the Fédérée. The Petroleum Division served the farm community with on-farm deliveries, but had an increasing presence in retail sales, with 163 Sonic service stations and 48 independents served by the Division. Operating revenues in petroleum departments totaled $217 million in 1988 and $232 million in 1989, which represented 13.5% of corporate revenues. In 1988, the

Division introduced the Sonic retail credit card. The Fruits and Vegetables Department sold only $3.8 million of produce, primarily as a reseller of imported potatoes.

The three significant marketing divisions of the Coopérative Fédérée were in livestock products. The Dairy Division generated 19.6% of corporate revenues in 1989 ($336 million) as a marketing agent for member cooperatives and in operating plants owned by the Division. The Division also had responsibility for participating in the provincial and national negotiations on quota allocation and pricing under the supply management regimes, and in provincial negotiations for hauling rates. The Dairy Division was affiliated with 8 regional dairy cooperatives, six of whom were the largest cooperatives in Québec, as measured by revenues. Table 2.1 shows the financial positions of these eight coops.

The Meat Division had sales of $308.8 million in 1989, a 2.5% decline from 1988. This level represented approximately 18% of Fédérée revenues in 1989 and was bolstered by $120.7 million in export sales. Exports of pork to the U.S. were approximately 80% of this volume, and represented nearly 22% of Canadian pork shipped to the U.S. The remainder was primarily delivered to Japan, which constituted 30% of Canadian pork exports there. The significant operating issue was a countervailing duty of 3.6 cents per pound levied by the U.S. in a frivolous trade action. The Division operated 3 slaughtering facilities and two distribution facilities in Québec. Two of the slaughter facilities killed both hogs and cattle.

The Poultry Division had sales of $321 million, or 18.7% of 1989 corporate revenues. The Division operated four slaughtering facilities, one plant for further processing, and a hatchery in Québec under the division name Bexel and one slaughter and processing facility in Ontario under the Galco label.

The Environment for Strategic Choice
in the Coopérative Fédérée

The Operating Environment and Strategic Choice

Organization theory attempts to describe the relationship between organizations and their environments. Among the competing theories that describe these relationships are several that require a deterministic role of the environment in the structure and behavior of individual organizations and their collective actions (Astley and Van de Ven, 1983; Peridis, 1989). These include the theories of population ecology (Hannan and Freeman,1977; Carroll, 1988) and industrial organization economics (Scherer, 1980), which consider the collective behavior of organizations within an industry; and contingency theory (Lawrence and Lorsch, 1967), which places individual firms within their

TABLE 2.1 Financial Position of Québec Daiiy Coops, 1988

Name	Net Sales	Total Assets	Members' Equity
Agropur	$ 795,772,000	$ 233,406,000	$ 105,664,000
Purdel	315,074,000	94,111,000	24,784,000
Agrinove	164,269,000	47,637,000	18,776,000
Chaîne Saguenay	75,537,000	25,553,000	7,746,000
Nutrinor	64,270,000	18,637,000	10,101,000
Cote Sud	61,163,000	32,434,000	10,617,000
Agrodor	35,401,000	9,183,000	3,426,000
St-Guillaume	15,832,000	4,820,000	1,765,000

Source: Annual Reports

environments. In these theories, organizational choices follow rather passively from the constraints of the operating environment (Astley and Van de Ven, 1983).

A contrasting view of the environment—organization interface places the voluntaristic actions of organizations in the center of analysis. Two schools of thought have developed with this perspective: strategic management (Child, 1972; Hofer and Schendel, 1978) and resource dependence (Pfeffer and Salancik, 1978). In these theories, the firm proactively seeks to improve its "fit" with the environment by restructuring its activities (organizational structure) or by negotiating relationships with other actors in the environment so as to mitigate the power of the operating environment to manipulate firm-level performance. In essence, the development of cooperatives as a form of business enterprise is an empirical example of what resource dependence theory predicts.

To be fair, organization theories exist that attempt to find common ground between these extreme views of environmental determinism. Porter's paper on the contributions of industrial organization theory to strategic management is one such bridge (Porter, 1985). Another is Peridis' concept of *environmental accommodation*, in which the degree of environmental determinism and organizational voluntarism depends on the various dimensions of the environment (Peridis, 1989). That is, some characteristics of the operating environment are deterministic or predictable, others unstable or ambiguous, and still others are unfathomably complex. Thus, organizations can proactivley (strategically) make choices to mitigate some of these factors, but cannot affect other characteristics.

To focus the discussion of the relationship of the Coopérative Fédérée to its environment, this paper will use the typology of task (operating) environments by Dess and Beard (1984). The task environment consists of the competitors, customers, suppliers, and regulatory bodies that directly affect organizational goal attainment (Bourgeois, 1980). The most common characterization of the task environment is the ubiquitous "five forces" model of Porter (1980) shown in Figure 2.5. Unfortunately, this model does not explicitly capture industry-specific government intervention as a factor determining industry attractiveness. The exclusion of this facet of the task environment in agriculture, particularly Canadian agriculture, is unfortunate, but does not diminish the value of inferences that can be drawn from the interactions of the firms that comprise the five forces. The Dess and Beard approach is to consider the description of the environment that comes from population ecology and resource dependence models of organization theory and to construct empirical measures of the dimensions of the environment that are consistent with SIC 4-digit industry classifications. Their environmental dimensions are not limited to Porter's model, but incorporate the effects of industry-specific governmental intervention. They collapse the description of the environment into three dimensions: *munificence, dynamism*, and *complexity*.

Munificence is the capacity or ability of the environment to sustain industry growth. This dimension is fundamental to the corporate portfolio models of strategic management, such as the BCG matrix. Munificence results from governmental intervention, decisions of rivals within the industry on R&D and technological investment, stage of product/industry life cycle, and on the actions of the other actors in Porter's model: supplier and buyer bargaining power, potential entrants, and substitutes.

Dynamism has two elements: instability and turbulence. Instability measures the rate of change and unpredictability of change. Wholey and Brittain (1989) characterize instability in three dimensions: frequency of change, amplitude of change, and predictability. In terms of organizational environment, the market for output (one resource on which the firm is dependent) will exhibit the results of these three characterizations of instability in terms of quarterly or annual changes in demand or price. Obviously, the greater the frequency, the larger the amplitude, and the higher the level of unpredictability, the greater is the overall instability of the industry environment. Turbulence refers to the interconnectedness among the organizations in the environment (Pfeffer and Salancik). That is, a turbulent environment for a given industry is characterized by the effects of decisions made by a large number of actors in the marketing channel. The vertically integrated U.S. broiler industry would be classified as significantly less turbulent than the Canadian industry, which has a multiplicity of government interventions and different firms at each level of the nonintegrated industry.

Complexity derives from firm-level characterization of a complex environment as one which results from heterogeneous activities of the firm. Thus, an industry with more inputs and outputs is more complex than an industry with firms producing few products or using few inputs. Likewise, an industry that which is dominated by diversified firms will be more complex than

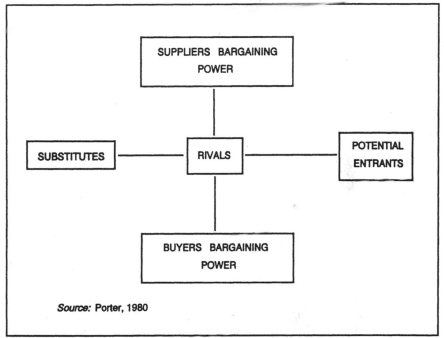

FIGURE 2.5 Porter's Model of the Task Environment

one dominated by specialized firms. A concentrated industry, particularly a geographically concentrated industry, is more complex because of the implied or actual jointness of the decisions taken by industry participants.

Strategic management and resource dependence models of organization—environment interface pose several normative responses to these dimensions of the environment. Among these implications are the following.

1. A munificent environment should encourage investment, entry, the building of barriers with excess resources to prevent entry, specialization of firms to exploit resources (inputs or markets) that are in abundance, and the search to expand organization scope.
2. An industry that lacks a munificent environment (illiberal) will provoke exit, restructuring and retrenchment, or diversification into other industries or technologies (seeking a more munificent set of inputs or markets).
3. A dynamic environment leads to building slack resources (excess capacity and uncommitted financial assets), collusion, long term contracts and vertical integration as strategies to buffer instability and the residual effects of decisions outside the control of industry rivals (turbulence).
4. A stable and nonturbulent environment requires significantly less buffering strategies such as holding slack resources, rent-seeking, and vertical relationships.

5. A complex environment implies decentralization and/or divisionalization of activities so as to develop specialized decision units to deal with tasks specific to markets for particular inputs or outputs.

6. An environment which is low in complexity (unconcentrated industry and with homogeneous activities among participating firms) supports strategies of centralization of decision making and efficiency-oriented firm structure.

The Task Environment for the Québec Dairy Industry

The task environment for the dairy industry had become increasingly illiberal and complex prior to the 1989 planning exercise for the Coopérative Fédérée. The low level of munificence can be attributed to (1) increasing rivalry for a market of fixed size in both the fluid and processing sub-industries, (2) a highly concentrated retail sector which demanded large shelf allocation fees, (3) an extraordinarily powerful supplier group (dairy farmers) which had become increasingly estranged from their cooperatives, and (4) a complex regime of government interventions in fluid and processed milk that limited the set of strategic choices available to the Fédérée and its constituent cooperatives. The high complexity was implied by the wide range of outputs from the industry which were variously affected by export opportunity and government regulation and by the extremely concentrated nature of the fluid and processing milk industries. The industry was turbulent in the sense that much of the strategic decision making was, because of dairy market regulation, vested in marketing boards, industry associations, and government agencies. Thus, much of the strategy of the dairy processing companies was dealing with choices made in the "residual environment" (Terreberry, 1968).

The task environment was not dynamic. The national and provincial marketing boards and supply management system for milk production limited price movements to forecastable or negotiated prices and quantities. Unpredictability was minimal. After the consolidation of the 1960s, there was little room for industry entry and exit was limited to small cheese processors and the closure of small, inefficient plants, rather than of significant competitors.

For the Coopérative Fédérée, the dairy sector of the Québec economy was of necessity the centerpiece of strategic analysis. The eight regionals who were members of the Fédérée had a large, though declining share of market in processed milk. Table 2.2 illustrates the recent history of the cooperative sector's share of milk received from pool 2, for processing into all products save fluid milk and cream. The coops had nearly the same market shares for fluid products. The significance of the pool 2 history is that in 1989, the national marketing board reduced the global quota for processing milk by 3% and Québec producers held 47% of national quota for pool two milk, (This cut was followed in 1990 and 1991 with additional cuts of 3% and 4.6%, respectively.)

TABLE 2.2 Market Share History for Pool 2 Milk (%)

Year	Cooperatives	Other Firms
1984-85	72.1	27.9
1985-86	69.7	30.3
1986-87	67.1	32.9
1987-88	65.4	34.6
1988-89	62.3	37.7
1989-90	60.0	40.0
1990-91	57.7	42.3

Hence, at the start of the planning exercise, the potential of processing industry overcapacity was known and a significant component of strategic analysis.

Overall, the dairy sector is the largest in Québec, with 36.7% of farm receipts. The combination of the size of the industry and the presence of the cooperatives in roughly 60% of the processing makes the Dairy Division strategic environment important. Three consumption trends were important to the task environment. Partially skimmed milk (2%) has been increasing as a percentage of consumption and whole milk has been declining. At the same time, butter consumption has continued its secular decline from 8.21 kilograms per capita in 1965 to 3.61 kg. per capita in 1989 (GREPA). Consumption of cheddar and specialty cheese had increased since 1980 (see Figure 2.6). These trends were important because (1) dairy farmers received payments for milk delivered to pool 1 (fluid) or to pool 2, depending on the type of quota they owned, *not a blend price*, and (2) an increasing amount of pool 1 milk was skimmed to make butter, which limited the market for pool 2 milk to be delivered for butter production. Thus, dairy farmers were affected by the consumer trend as the price differential between pool 1 and pool 2 prices increased to $1.33 per hectolitre in 1988-89 (GREPA). This caused great dissension between the two groups of producers.

The dairy products market in Canada is organized in two distinct sub-industries, fluid and processing. The fluid market is controlled by the provinces; each province oversees a production quota system designed to meet provincial needs in the fluid market. Fluid milk does not cross provincial borders. In Québec, the Régie de marchés agricoles du Québec (RMAQ) is the marketing board with powers to oversee negotiations between the milk producers' union, the Fédération des producteurs de lait du Québec (FPLQ) and two processors' groups, the noncooperative sector (Conseil de l'industrie laitière) and the cooperatives (Conseil des coopératives laitières). These negotiations lead to setting prices for raw milk in the six milk classes that form the two pools, as well as the terms of delivery, levies, and transportation. In the fluid sector, the RMAQ also sets wholesale and retail fluid milk prices.

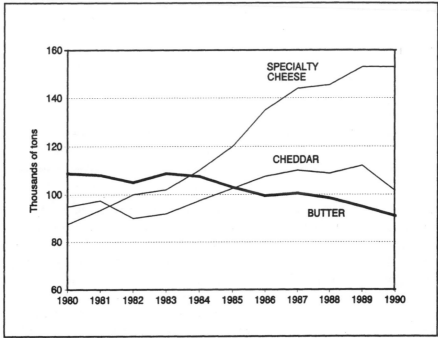

FIGURE 2.6 Consumption of Butter and Cheese, Québec

The industrial sector has the additional complexity of the quota system deriving from national supply management. The Canadian Dairy Council is the organization that oversees the national system, setting support prices for butter, cheddar, and milk powder and the farm level subsidy. The CDC and the provincial producer's organizations, such as the FPLQ, constitute the federal marketing board, which sets global and provincial quotas for production of industrial milk products on a butterfat basis.

Figure 2.7 shows the price effects of the supply-managed marketing board system, relative to U.S. prices. If the GATT round causes the border protections against imports of manufactured dairy products to disappear, then the Québec dairy sector will face global competition. This will further limit the cooperatives' abilities to maintain operating margins given the pricing of raw milk.

The consequence of this highly regimented marketing regime has been the estrangement of the farmer-members from their coops. Rather than using the cooperatives as an organization for mutual gain through market power, the milk producers use their federation with its attendant legal power. Opposite the federation at every negotiation is the Coopérative Fédérée and the 8 member cooperatives. The regulations of the marketing boards include the rights of all the plants to raw milk supply, regardless of coop—member relationships.

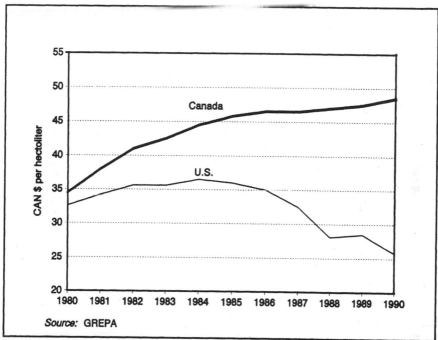

FIGURE 2.7 Milk Support Prices in Canada and the U.S.

Eventually, the relationship between the coop member and the coop becomes virtually nonexistent. In 1988-89, the dairy cooperatives processed 8.6% less milk than their members produced; the difference went to non-cooperative plants (Côté and Vézina, 1989).

Côté and Vézina suggest that dairy cooperatives will mutate into something that looks like investor-owned firms as a result of this task environment. Regardless if this comes to pass, the situation in 1988-89 required the Coopérative Fédérée to make a change, to protect the market position of its constituent dairy cooperatives, if not their farmer-members.

The Task Environment for the Pork and Broiler Industries

Compared to the dairy industry, the task environments for the other two marketing divisions of the Coopérative Fédérée are less complex and more munificent. While the poultry industry is also supply-managed, there is more strategic flexibility. Prices are negotiated with growers, but there are no support prices or regulated retail prices to compress the strategic decision set. As well, the Canadian poultry industry is more fragmented than the dairy industry, without the core of oligopolistic processors. The Bexel Division competed in 1988 with Tyson Canada, several small primary and further processors, and was protected behind the import restrictions of the national supply management program.

The rigidities of the poultry supply management system cause supply—demand imbalances and costly inventories from time to time. It is difficult to obtain broilers for new product development, such as value-added (further) processing. The supply management system maintains the identities of the various levels of the market and effectively prevents the manner of integration existent in the U.S. To compete effectively in a world market or an integrated North American market, if the GATT round eliminates the import quotas that protect supply management, would require horizontal scale and additional investment in cost-saving technology.

The pork market in 1988 was, like cereal grains, a North American market. Turcotte et Turmel exported 30% or more of their annual volume to the eastern United States. The competition in Quebec consisted of a significant rival in Olympia and a group of small, specialized hog slaughtering firms of increasing total market share. Otherwise, the principal rivals for Turcotte et Turmel were American firms that competed for the eastern U.S. markets. In addition, there was significant excess capacity among the slaughter facilities which placed cost pressures on Turcotte et Turmel, which was a high cost competitor due to age of plants and the nonspecialization of most of the plants. Changeover costs for beef/pork plants were high (Le Coopérateur Agricole, June 1992). Thus, the task environment was not highly munificent.

The pork industry in Québec exhibited the dynamism associated with the open North American market. The volatility of U.S. production and prices translated into volatility in Québec. In addition, the U.S. had placed countervailing duties on hogs and pork from Canada, based upon the contention that the tripartite stabilization scheme (producers, provincial government, federal government) constituted a trade subsidy. This unpredictable facet of the task environment caused even greater dynamism.

Strategic Choices in Reconfiguring the Coopérative Fédérée

During the 1990 fiscal year ending in November, the Coopérative Fédérée made several changes. First, it sold all of its assets in the Dairy Division to a limited partnership (société en commandite) of five dairy processors: Purdel, Agrinove, Agridor, Nutrinor, and Côte Sud (see Table 2.1). The limited partnership, Lactel, produces only manufactured products and represents a restructuring of the non-fluid milk operations of the five coops and the Coopérative Fédérée. Lactel was unable to attract the largest of the dairy coops, Agropur, to membership. This may have been due to the nature of governance of the partnership. Each of the six principals has two seats on the board, except Agrinove, which was the most heavily involved in manufacturing milk and received three. Agropur, larger than all of the other principals combined, could not involve more than 25% of the seats according to the partnership agreement (Le Coopérateur Agricole, August 1990).

The choice of a limited partnership was made to separate voice from claim on earnings. While net earnings (or losses) are partitioned by net investment, control is accorded to the board members by the formula described above. This structure is different from both the normal corporate governance and from federated cooperatives. The limited partnership structure prevents double taxation of earnings, required by law of corporations, and to a large degree, of federated cooperatives.

The transaction showed a disposal by the Coopérative Fédérée of:

Working capital	$ 46,981,000
Investments	1,312,000
Fixed assets	344,000
Goodwill	9,000,000
	$ 57,637,000

In return, the company received:

Notes receivable	$ 28,637,000
Common shares	13,000,000
Preferred shares	16,000,000
	$ 57,637,000

The valuation of assets netted an extraordinary gain of $6,795,677 in 1990.

The Coopérative Fédérée also invested approximately $8 million in other assets and participations, including the purchase of a competitor in the meat industry, Les Abattoirs R. Roy; majority interest in a slaughterhouse in Ontario, Saint-Isadore de Prescotte; and minority interest in a slaughter facility in Harlan, Iowa. In the heretofore neglected horticulture sector, the company purchased Les Patates Québécoises, a potato processing firm, for $7,201,000.

The Fédérée continued to make changes in 1991. The two significant additions to the marketing divisions were (1) purchasing the shares of Tyson Canada for $55,139,000, and (2) the consolidating of Groupe Olympia with Turcotte et Turmel in the Meat Division. The Tyson purchase placed the resultant Poultry Division as the largest poultry processor in Canada, with 60% of the processing capacity in Québec (Service de l'information et des statistiques, 1992). The Meat Division, renamed Olymel, is organized as a limited partnership. The operating results are reported in the consolidated results of the Coopérative Fédérée, although the Co-op owns only 50% of the shares in the partnership. Olymel controls approximately 70% of the total slaughter capacity in the province for pork, although it has begun a campaign of rationalization to close and remodel the regrouped assets.

In addition to the changes in the marketing divisions, the Coopérative Fédérée also purchased 23 Esso stations for the Petroleum Division and 50% of a grain and fertilizer distribution facility. The Coopérative Fédérée also committed itself to decentralizing the Farm Supply Group, forming autonomous coops from retail assets. A planning group was organized to assist management of farm supply locals to improve and control performance, while maintaining autonomy from the federation.

Currently, Lactel is ending its first year of operation. Lactel management have joint control over a significant portion of the assets in manufactured dairy products. This has allowed them to close three manufacturing facilities and 6 milk assembly facilities. It has also permitted the change from butter operations to cheese operations, which are more profitable and fill a growing market. Arguably, the individual cooperatives could not have made these choices. In the long term, Lactel should be able to lead a faster redeployment of assets than the constituent partners could have.

Unfortunately, in the short term, these choices are costly. The Coopérative Fédérée's share of the first year losses of Lactel (1991) was $1.8 million. In the next few years, the competition with Agropur and the noncooperative sector may cause Lactel to lose additional money. It is difficult to speculate, as the magnitude of change possible from the GATT round is potentially enormous.

There are no results yet from the investments in the Poultry and Meat Divisions. Both divisions are in the process of rationalizing assets and attaining a consistent culture among the new and old groups of employees in the divisions.

However, it is clear that the divisionalization of the Coopérative Fédérée places it in a better position to deal with the differences in the task environments of its businesses. In the illiberal dairy industry, retrenchment and restructuring of assets is a response predicted by organization theory. The development of Lactel as the specialized decision unit in manufactured dairy products is also predicted as a response to the complexity of the task environment. The dynamic pork industry, with its limited munificence due to overcapacity and reliance on exports to the U.S., caused the Coopérative Fédérée restructure its assets into Olymel to better control overhead costs and to minimize the effects of dynamism by holding slack resources. In the Poultry Division, the Coop has responded to the task environment by purchasing additional assets to optimize in this stable, low-complexity industry. It only remains to measure the efficacy of these choices of the relevant strategic time horizon.

Notes

1. The section on the history is based upon *l'Histoire de la coopération agricole et la Coopérative Fédéréé de 1922 à nos jours*, an unpublished manuscript from the archives of the company.

References

Astley, W.G., and A.H. Van de Ven. 1983. Central Perspectives and Debates in Organization Theory. *Administrative Science Quarterly*. 18: 245-273.

Bourgeois, L.J. 1980. Strategy and Environment: A Conceptual Integration. *Academy of Management Review*. 5(1):25-39.

Carroll, G.R. 1988. *Ecological Models of Organizations.* Cambridge, MA: Ballinger.

Child, J. 1982. Organization Structure, Environment and Performance: The Role of Strategic Choice. *Sociology*. 6: 1-22.

Le Coopérateur Agricole. Montréal: Coopérative Fédérée de Québec. various issues.

Coopérative Fédérée de Québec. Annual Reports. 1980-1991.

Coopérative Fédérée de Québec. *l'Histoire de la Coopération et la Coopérative Fédérée de 1922 à nos jours*.

Côté, D. and M. Vézina. 1989. *La mutation de l'entreprise coopérative: Le cas de l'industrie laitière québécoise*. Montréal: École des hautes études commerciales de Montréal. Cahier 89-2.

Dess, G.G. and D.W. Beard. 1984. Dimensions of Organizational Task Environments. *Administrative Science Quarterly*. 29: 52-73.

GREPA (Groupe de recherche en économie et politiques agricoles). 1991. *Québec Dairy Facts*. Québec: Université Laval.

Hannan, M. and J. Freeman. 1877. The Population Ecology of Organizations. *American Journal of Sociology*. 82: 929-964.

Hofer, C.W. and D. Schendel. 1978. *Strategy Formulation: Analytical Concepts*. St. Paul: West Publishing.

Lawrence, P. and J. Lorsch. 1967. *Organization and Environment*. Boston: Harvard Business School.

Peridis, T. 1989. Task Environment and Organizational Diversity. Paper presented at 1989 Academy of Management annual meeting.

Pfeffer, J. and G. Salancik. 1978. *The External Control of Organizations: A Resource Dependence Perspective*. New York: Harper and Row.

Porter, M. 1980. *Competitive Strategy*. New York: Free Press.

Scherer, F.M. 1980. *Industrial Market Structure and Economic Performance, 2nd edition*. Chicago: Rand McNally.

Service de l'information et des statistiques. 1992. *L'industrie bioalimentaire au Québec: Bilan 1991 et Perspectives*. Ministère de l'Agriculture, des Pêcheries et de l'Alimentation du Québec.

Terreberry, S. 1968. The Evolution of Organizational Environments. *Administrative Science Quarterly*. 12: 590-613.

Wholey, D.R. and J. Brittain. 1989. Characterizing Environmental Variation. *Academy of Management Journal*. 32(4):867-882.

## PART TWO

Strategic Marketing: Integration and Differentiation

3

Market Incentives for Cooperative Forward Integration into Processing Activities

Jeffrey S. Royer and Sanjib Bhuyan

Farmer cooperatives are typically involved in first-stage marketing and food processing activities as a result of their role as vertical extensions of the farming operations of their members. Consequently, the marketing and processing activities in which farmer cooperatives participate are usually characterized by little market power and low margins (Rogers and Marion, 1990). Considerable discussion has focused on explaining why cooperatives have not integrated forward into high-margin, value-added activities to a greater extent. Explanations include arguments that: (1) the production orientation of directors restricts the ability of a cooperative board to supervise and assist management as the organization's scope grows vertically and increasingly involves consumer-oriented merchandising activities (Jamison, 1960), (2) cooperatives are disadvantaged by scale economies associated with complex organizational tasks (Caves and Petersen, 1986), and (3) cooperatives are often insufficiently capitalized to make the substantial investments in research and development and in advertising that are necessary to be successful in processed markets (Rogers and Marion, 1986).

Unfortunately, there has been little theoretical analysis of the incentives cooperatives may have for integrating forward into later processing stages despite its importance to cooperatives and their members. Perhaps the most relevant work has been that of Eisenstat and Masson (1978), who concluded that in the situations where vertical integration by cooperatives could provide producers and consumers the greatest benefits, there may not be an incentive for cooperatives to integrate (p. 51). In this paper, we develop a model for evaluating the incentives cooperatives have for integrating forward from marketing into processing activities within the framework of a bilateral monopoly. Although we choose this structure because it is the simplest with which to begin, it also has considerable appeal due to its prevalence in some cooperative markets, particularly markets dealing in perishable commodities such as milk, fruits, and vegetables. We analyze vertical integration by cooperatives under two sets of behavioral assumptions. In addition, we analyze integration by an investor-owned firm (IOF) as a basis for comparing the incentives cooperatives have to integrate forward and the price and quantity effects that would follow. Our results provide an additional explanation, based on market power, for the relatively low degree of integration by cooperatives.

Theoretical Basis

Our analysis is based on theoretical concepts in three areas: (1) bilateral monopoly, (2) the determinants of vertical integration, and (3) cooperatives. Despite the attention given to bilateral monopoly in industrial organization literature, little analysis has focused on bilateral monopoly in the context of vertical integration decisions. Many of the quantitative vertical integration models are based on the concept of successive monopoly (Sheldon 1991, p. 8). These models assume "arm's length" pricing by the upstream monopoly, i.e., the downstream monopoly (or monopolies) accepts the price as parametric and exercises no monopsony power. Thus, models based on this assumption ignore the possibility of dominance by the downstream monopoly-monopsony in the bilateral case. In our model, we consider dominance by both the monopolist and the monopsonist.

Three broad determinants of vertical integration are commonly cited: (1) technological economies, (2) transactional economies, and (3) market imperfections (Perry 1989, p. 187). Technological economies of integration are based on physical interdependencies in the production process. The usual example is the heating and handling economies that lead to integration in the production of iron and steel. Transaction costs are associated with the process of exchange instead of production. In some situations, the market may fail as an efficient means of coordinating economic activity. As a result, a firm may be able to reduce its transaction costs by integrating. For example, in the case of a bilateral monopoly, either firm may be able to eliminate the costs of negotiating and enforcing a contract with the other through integration.

Market imperfections that may produce incentives for vertical integration include imperfect competition in addition to imperfections caused by externalities and imperfect or asymmetric information. As Perry observes (p. 189), market imperfections are an important determinant of vertical integration. Vertical integration in response to technological or transactional economies can be expected to increase economic welfare. Thus the primary focus of transaction cost economics is explaining and predicting patterns of vertical integration. On the other hand, vertical integration in reaction to market imperfections may either increase or decrease welfare. As a result, public policy questions become important.

Because we have no *a priori* reasons to assume that the technological or transactional incentives for cooperatives to integrate forward differ from those of other firms, we will focus on market incentives. Specifically, this paper examines the incentives for integration that arise from the ability of the integrated firm to maximize the aggregate profits of the bilateral market structure in contrast to both firms independently maximizing individual profits without taking into account the incremental profit of the other.

In the classic Helmberger and Hoos model of a marketing cooperative, the objective of the cooperative is to maximize the raw product price for whatever quantity producers choose to supply. Consistent with this objective, the short-run equilibrium for a cooperative assembler and its producers would occur where the average net return (defined here as average revenue from the assembled product less the per-unit handling cost) equals the supply price. At equilibrium, the cooperative breaks even because payments to producers exhaust cooperative surplus. In contrast, a cooperative that attempts to maximize producer welfare would set the raw product price equal to the marginal net return (marginal revenue from the assembled product less the per-unit handling cost). At this level of output, the cooperative would earn a surplus, which would be distributed to producers as a patronage refund. Several authors (for example, see Cotterill 1987, pp. 190-92; Schmiesing 1989, pp. 159-62; or Staatz 1989, pp. 4-5) have suggested that the cooperative would be unsuccessful in restricting producer output to this level. They argue that at any level of output less than the breakeven equilibrium level, the receipt of patronage refunds will provide producers an incentive to expand output until it reaches the quantity at which the average net return equals the supply price.

Variable-Proportions Model

We consider a bilateral monopoly consisting of an assembler and a processor. Producers (A) sell a single raw product q_A to the assembler (B), which markets the product q_B (which is simply q_A assembled) to the processor (C). The processor uses q_B in manufacturing a finished product q_C, which it sells to consumers. The finished product q_C is manufactured from q_B and another factor w according to a variable-proportions production function with the usual properties. We assume that the assembler faces an upward-sloping raw product supply curve and the processor faces a downward-sloping final product demand curve. The assembler's cost of handling the raw product is assumed to be constant. Factor w is supplied competitively at a constant price r.

The assembler is alternately assumed to be an IOF and a cooperative. Both the IOF and cooperative analyses are conducted for both the case of assembler dominance and the case of processor dominance. Although the price paid the assembler by the processor will depend on the relative bargaining power of the two parties, these solutions are useful in identifying the bounds for the price and quantity outcomes. The cooperative analyses are conducted under two alternate behavioral assumptions. Under the first, the cooperative maximizes the joint profits of producers and the assembler by successfully satisfying the appropriate first-order condition. Under the second, the cooperative is passive in that it does not or cannot set the quantity of raw product it handles in order to maximize joint profits. Instead it accepts whatever quantity of output producers choose to market. The receipt of patronage refunds provides producers an incentive to expand output until the cooperative breaks even, as in the Helmberger and Hoos model.

IOF Assembler: Assembler Dominance

The processor's profit function is

$$\pi_C = p_C q_C(p_B, w) - p_B q_B - rw \tag{1}$$

where p_i ($i = A$, B, or C) represents the price received by the producers, assembler, or processor. The processor exercises monopoly power in the final product market. However, because the assembler is dominant, the processor takes the price the assembler sets for the assembled raw product. Thus the processor's first-order conditions for profit maximization are

$$\frac{\partial \pi_C}{\partial q_B} = \left(p_C + q_C \frac{dp_C}{dq_C} \right) \frac{\partial q_C}{\partial q_B} - p_B = 0$$

$$\frac{\partial \pi_C}{\partial w} = \left(p_C + q_C \frac{dp_C}{dq_C} \right) \frac{\partial q_C}{\partial w} - r = 0. \tag{2}$$

These conditions imply that the processor should use each factor at the level at which its marginal revenue product equals its price. The simultaneous solution of (2) for q_B and w yields the derived demand functions for the inputs:

$$q_B = q_B^* (p_B, r, p_C)$$

$$w = w^* (p_B, r, p_C). \tag{3}$$

The assembler is a monopsonist in the raw product market and a monopolist in the assembled product market. Its profit function is

$$\pi_B = P_B q_B - P_A q_A - h q_B \tag{4}$$

where h is the per-unit handling cost. Because the assembler is dominant, it sets the price for the assembled product in order to maximize profit given the processor's demand function. Recalling that $q_B = q_A$ and substituting the processor's demand function for q_B from (3) into (4), the assembler's first-order condition for profit maximization is derived:

$$\frac{d\pi_B}{dq_B} = \left(P_B + q_B \frac{dp_B}{dq_B}\right) - \left(P_A + q_A \frac{dp_A}{dq_A}\right) - h = 0. \tag{5}$$

This condition can be rewritten

$$MFC_A + h = MR_B. \tag{6}$$

The dominant assembler maximizes its profit when its marginal factor cost plus the per-unit handling cost equals its marginal revenue.

IOF Assembler: Processor Dominance

The assembler's profit function is (4). Because the processor is dominant, the assembler takes the price it receives for the assembled product as given while exercising monopsony power in the raw product market. Its first-order condition is

$$\frac{d\pi_B}{dq_B} = P_B - \left(P_A + q_A \frac{dp_A}{dq_A}\right) - h = 0. \tag{7}$$

From this, we derive the assembler's inverse factor supply function:

$$P_B = MFC_A + h. \tag{8}$$

Substituting (8) into (1), the processor's profit function is

$$\pi_C = P_C q_C(p_B, w) - (MFC_A + h)q_B - rw. \tag{9}$$

The corresponding first-order conditions are

$$\frac{\partial \pi_C}{\partial q_B} = \left(P_C + q_C \frac{dp_C}{dq_C} \right) \frac{\partial q_C}{\partial q_B} - \frac{d(MFC_A q_A)}{dq_A} - h = 0$$

$$\frac{\partial \pi_C}{\partial w} = \left(P_C + q_C \frac{dp_C}{dq_C} \right) \frac{\partial q_C}{\partial w} - r = 0. \tag{10}$$

The first condition in (10) can be rewritten

$$\frac{d(MFC_A q_A)}{dq_A} + h = MRP_B. \tag{11}$$

The dominant processor maximizes its profit when the value marginal to the assembler's marginal factor cost function plus the per-unit handling cost equals the marginal revenue product of the assembled raw product.

IOF Assembler: Post-Integration

If the assembler integrates forward by acquiring the processor, it will maximize the joint profits from assembling and processing the raw product:

$$\pi_{BC} = P_C q_C(q_B, w) - P_A q_A - h q_B - rw. \tag{12}$$

The new firm is a monopsonist in the raw product market and a monopolist in the final product market. The first-order conditions for profit maximization are

$$\frac{\partial \pi_{BC}}{\partial q_B} = \left(P_C + q_C \frac{dp_C}{dq_C} \right) \frac{\partial q_C}{\partial q_B} - \left(P_A + q_A \frac{dp_A}{dq_A} \right) - h = 0$$

$$\frac{\partial \pi_{BC}}{\partial w} = \left(P_C + q_C \frac{dp_C}{dq_C} \right) \frac{\partial q_C}{\partial w} - r = 0. \tag{13}$$

The first condition in (13) can be rewritten

$$MFC_A + h = MRP_B. \qquad (14)$$

The integrated firm maximizes its profit when its marginal factor cost plus the per-unit handling cost equals the marginal revenue product of the assembled raw product.

Active Cooperative Assembler: Assembler Dominance

Here we assume that the cooperative maximizes the joint profits of producers and the assembler:

$$\pi_{AB} = p_B q_B - F - h q_A \qquad (15)$$

where F is total on-farm production costs. Substituting the processor's demand function for q_B from (3), the first-order condition for profit maximization is

$$\frac{d\pi_{AB}}{dq_B} = \left(p_B + q_B \frac{dp_B}{dq_B} \right) - \frac{dF}{dq_A} - h = 0 \qquad (16)$$

and can be rewritten

$$MC_A + h = MR_B. \qquad (17)$$

The cooperative maximizes the joint profits of producers and the assembler when the marginal cost of producing the raw product plus the per-unit handling cost equals the marginal revenue from the assembled raw product.

Active Cooperative Assembler: Processor Dominance

If the processor is dominant, the cooperative assembler takes the price set by the processor for the assembled product. Thus the first-order condition for maximization of the cooperative's objective function (15) is

$$\frac{d\pi_{AB}}{dq_B} = p_B - \frac{dF}{dq_A} - h = 0. \qquad (18)$$

Rearranging, we derive the cooperative's inverse factor supply function:

$$p_B = MC_A + h. \qquad (19)$$

Substituting this into the processor's profit function (1), the latter is rewritten

$$\pi_C = P_C q_C(p_B, w) - (MC_A + h)q_B - rw. \tag{20}$$

The first-order conditions for profit maximization are

$$\frac{\partial \pi_C}{\partial q_B} = \left(P_C + q_C \frac{dp_C}{dq_C} \right) \frac{\partial q_C}{\partial q_B} - \frac{d(MC_A q_A)}{dq_A} - h = 0$$

$$\frac{\partial \pi_C}{\partial w} = \left(P_C + q_C \frac{dp_C}{dq_C} \right) \frac{\partial q_C}{\partial w} - r = 0. \tag{21}$$

Assuming producers attempt to maximize their profits by setting the marginal cost of producing the raw product to the price they receive, the first condition in (21) can be rewritten

$$MFC_A + h = MRP_B. \tag{22}$$

The dominant processor maximizes its profit when the assembler's marginal factor cost plus the per-unit handling cost equals the marginal revenue product of the assembled raw product.

Active Cooperative Assembler: Post-Integration

If the cooperative assembler integrates forward by acquiring the processor, it will maximize the joint profits from producing, assembling, and processing the raw product:

$$\pi_{ABC} = P_C q_C(q_B, w) - F - hq_B - rw. \tag{23}$$

The first-order conditions for this objective are

$$\frac{\partial \pi_{ABC}}{\partial q_B} = \left(P_C + q_C \frac{dp_C}{dq_C} \right) \frac{\partial q_C}{\partial q_B} - \frac{dF}{dq_A} - h = 0$$

$$\frac{\partial \pi_{ABC}}{\partial w} = \left(P_C + q_C \frac{dp_C}{dq_C} \right) \frac{\partial q_C}{\partial w} - r = 0. \tag{24}$$

The first condition in (24) can be rewritten

$$MC_A + h = MRP_B. \tag{25}$$

The cooperative maximizes the joint profits from producing, assembling, and processing the raw product when the marginal cost of producing the raw product plus the per-unit handling cost equals the marginal revenue product of the assembled raw product.

Passive Cooperative Assembler: Assembler Dominance

Here we assume that the cooperative is passive in terms of accepting whatever quantity of raw product producers choose to market. Producers recognize the existence of patronage refunds and produce the quantity for which marginal cost equals the sum of the price and the per-unit patronage refund:

$$MC_A = p_A + s. \tag{26}$$

The per-unit patronage refund equals the profit of the cooperative assembler divided by the quantity of raw product assembled:

$$s = \frac{p_B q_B - p_A q_A - hq_B}{q_A} \tag{27}$$

$$= p_B - p_A - h.$$

Substituting (27) into (26), we derive the cooperative's inverse factor supply function, which is identical to (19) for the active cooperative assembler under processor dominance. Solution of the first-order conditions in (2) for p_B yields the inverse demand function of the processor for the assembled raw product:

$$p_B = p_B^*(q_B, r, p_C). \tag{28}$$

Setting the cooperative's inverse factor supply function (19) equal to the processor's inverse factor demand function (28), we derive the equilibrium solution:

$$p_B^* = MC_A + h. \tag{29}$$

Equilibrium occurs at the quantity for which the price of the raw product equals the marginal cost of producing and assembling it.

Passive Cooperative Assembler: Processor Dominance

Solution of the model is identical for a dominant processor regardless of whether it purchases the assembled product from a cooperative actively pursuing the joint profit function (15) or one that passively accepts whatever quantity of raw product producers choose to market. After substituting the cooperative's inverse factor supply function (19) into the processor's profit function (1), the latter is equivalent to (20), the profit function of a dominant processor that purchases from an active cooperative. The first-order conditions are equivalent to those in (21).

Passive Cooperative Assembler: Post-Integration

If the passive cooperative assembler integrates forward by acquiring the processor, it will still accept whatever quantity of raw product producers choose to market. Producers again determine the quantity of raw product according to (26). However, the per-unit patronage refund is now

$$s = \frac{p_C q_C - p_A q_A - h q_B - rw}{q_A}$$

(30)

$$= \frac{p_C q_C}{q_A} - p_A - h - \frac{rw}{q_A}.$$

Substituting (30) into (26) and rearranging, we derive the equilibrium solution:

$$MC_A + h = \frac{p_C q_C - rw}{q_A}$$

(31)

where the right-hand side is the cooperative's average net return. Equilibrium occurs at the quantity for which the marginal cost of producing and assembling the raw product equals the cooperative's average net return from the processed product.

Summary of Solution Conditions

A summary of the solution conditions is presented in Table 3.1 for IOF, active cooperative, and passive cooperative assemblers under variable-proportions production technology. Examination of these conditions reveals some important differences in the behavior of the three types of firms:

1. Both active and passive cooperative assemblers behave like competitive

TABLE 3.1 Summary of Solution Conditions for Variable-Proportions Technology

	Investor-Owned Firm	Active Cooperative	Passive Cooperative
Assembler Dominance	$MFC_A + h = MR_B$	$MC_A + h = MR_B$	$MC_A + h = p_B$
Processor Dominance	$\dfrac{d(MFC_A q_A)}{dq_A} + h = MRP_B$	$MFC_A + h = MRP_B$	$MFC_A + h = MRP_B$
Post-Integration	$MFC_A + h = MRP_B$	$MC_A + h = MRP_B$	$MC_A + h = \dfrac{p_C q_C - rw}{q_A}$

firms in the raw product market whereas the IOF assembler exercises monopsony power.

2. The passive cooperative assembler behaves like a competitive firm in the assembled and final product markets whereas both the IOF and active cooperative assemblers exercise monopoly power.

The exercise of monopsony power by the assembler is indicated by the existence of MFC_A instead of MC_A on the cost side (left-hand side) of the assembler-dominant and post-integration conditions. Monopoly power is indicated by MR_B on the revenue side of the assembler-dominant conditions and MRP_B on the revenue side of the post-integration conditions.

The passive cooperative assembler's behavior is similar to that of a competitive firm because of p_B in the assembler-dominant condition and $(p_C q_C - rw)/q_A$, or average net return, in the post-integration condition. The average net return can be related to the marginal revenue product of the assembled raw product in the post-integration conditions for the IOF and active cooperative assemblers insofar as *marginal* net return can be defined as

$$\frac{\partial(p_C q_C - rw)}{\partial q_A} = \left(p_C + q_C \frac{dp_C}{dq_C}\right)\frac{\partial q_C}{\partial q_B} \tag{32}$$

where we recognize the right-hand side as $MR_C \cdot MPP_B$ or MRP_B.

The processor exercises monopsony power in the assembled product market, when dominant, and monopoly power in the final product market. The terms $d(MFC_A q_A)/dq_A$ and MFC_A in the processor-dominant conditions result from marginalization of the assembler inverse factor supply functions.[1] Because the supply functions for the active and passive cooperatives are identical, the processor-dominant solution conditions for the cooperatives are also the same.

Notice that the solution conditions for the active and passive cooperatives under processor dominance are equivalent to that for the IOF after integration. In addition, the condition for the passive cooperative under assembler dominance is equivalent to that for the active cooperative after integration because the processor purchasing from the passive cooperative will set the marginal revenue product of the assembled raw product equal to its price according to the first condition in (2).

Fixed-Proportions Model

A fixed-proportions production function may be appropriate for representing the technology used in processing many agricultural commodities. Although within a certain range, additional capital and labor might increase the technical efficiency with which raw product is converted into final product by reducing waste and spoilage, these factors cannot be generally substituted for raw product in order to increase final product output. An advantage of assuming a fixed-proportions technology is that it lends itself more readily to graphical exposition.

We have also developed a model based on a fixed-proportions relationship between the raw and final products, which is the subject of another paper (Bhuyan and Royer, 1992). In that model, we assume that the processor's cost of finishing the raw product is constant. For convenience and without loss of generality, we also assume that the processor produces one unit of final product from each unit of raw product. As a result, we can dispense of subscripts on the quantities, and the processor's profit function can be written

$$\pi_C = p_C q - p_B q - kq \tag{33}$$

where k is the per-unit processing cost. Although we do not present the derivation of the fixed-proportions model here, the solution conditions are summarized in Table 3.2.

The assumptions of the fixed-proportions model allow us to work with more explicit factor demand and supply functions, thus eliminating some of the price, marginal revenue, and marginal revenue product terms in the solution conditions. They also allow us to see explicitly the double marginalization in the assembler-dominant conditions, as represented by the existence of the term $d(MR_C q)/dq$ on the revenue side of the conditions for the IOF and active cooperative assemblers. Again, we see that the solution conditions for the active and passive cooperative assemblers under processor dominance are equivalent to that for the IOF assembler after integration and that the condition for the passive cooperative under assembler dominance is equivalent to that for the active cooperative after integration.

The solutions in Table 3.2 are compared graphically in Figure 3.1, where $MMFC_A$ and MMR_C respectively represent $d(MFC_A q)/dq$ and $d(MR_C q)/dq$, i.e., the schedules marginal to the marginal factor cost and marginal revenue curves. Points 1 and 2 represent the solutions for the IOF assembler under the respective conditions of assembler and processor dominance. Point 3 represents the solution for the IOF after integration. The post-integration solution is characterized by greater output and a lower final product price than either of the pre-integration solutions. Thus consumers are better off as a result of integration by the IOF. The price paid producers is read from the raw product supply curve, found by subtracting the constant handling and processing costs from $MC_A + h + k$. Integration increases the price paid producers in addition to increasing output. Thus producers are also better off because of integration by the IOF.

Point 4 represents the solution for the active cooperative under assembler dominance. Output and raw product price are greater and final product price

TABLE 3.2 Summary of Solution Conditions for Fixed-Proportions Technology

	Investor-Owned Firm	Active Cooperative	Passive Cooperative
Assembler Dominance	$MFC_A + h = \dfrac{d(MR_C q)}{dq} - k$	$MC_A + h = \dfrac{d(MR_C q)}{dq} - k$	$MC_A + h = MR_C - k$
Processor Dominance	$\dfrac{d(MFC_A q)}{dq} + h = MR_C - k$	$MFC_A + h = MR_C - k$	$MFC_A + h = MR_C - k$
Post-Integration	$MFC_A + h = MR_C - k$	$MC_A + h = MR_C - k$	$MC_A + h = p_C - k$

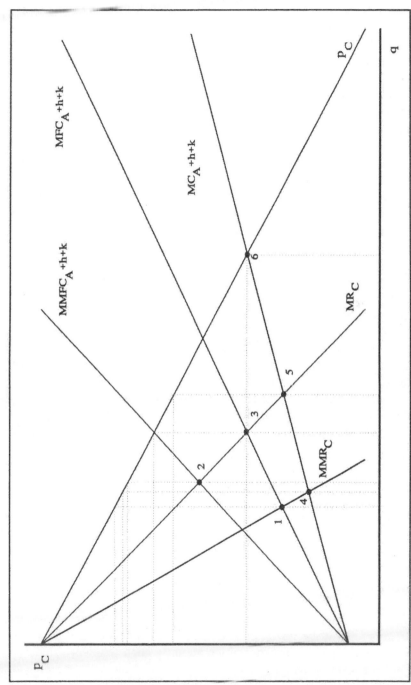

FIGURE 3.1 Price and Output Under Alternative Vertical Market Structures and Fixed-Proportions Production Technology

is less than when the assembler is a dominant IOF. Thus, under assembler dominance, both consumers and producers are better off when the assembler is an active cooperative, irrespective of the payment of patronage refunds to producers. The solution for the active cooperative under processor dominance is point 3, identical to that for the IOF after integration. Thus, an active cooperative under processor dominance provides consumers and producers the same benefits as an integrated IOF without integrating. However, by integrating, an active cooperative can further improve the situation of both consumers and producers, irrespective of patronage refunds, as indicated by the prices and output associated with point 5. This solution also represents an improvement over the unintegrated active cooperative under assembler dominance.

The solution for the passive cooperative under assembler dominance is also represented by point 5. Thus, under assembler dominance, a passive cooperative provides the same benefits to consumers as an integrated active cooperative without integrating. However, producers will be better off with the integrated cooperative because any patronage refunds they receive will include the profits of the processing plant. Note that the solution for the passive cooperative under assembler dominance yields a greater output and raw product price and a lower final product price than the passive cooperative under processor dominance, which is represented by point 3. In the case of IOF and active cooperative assemblers, no generalizations can be made about the comparative output and prices under assembler and processor dominance. Whether output will be greater or less under processor dominance will depend on the specific demand and cost functions. However, the output of a passive cooperative will be less under processor dominance because the inability of the cooperative to set the quantity of raw product it handles is replaced by the discipline of the processor, which is a monopsonist in the assembled product market. Producers will be worse off than under assembler dominance because the net price paid producers (the cash price plus the per-unit patronage refund), which is read from the raw product supply curve in the case of a passive cooperative, will also be less. The solution for a passive cooperative under processor dominance is identical to those for the IOF after integration and the active cooperative under processor dominance. Thus, when the processor is dominant, the output and price results are the same regardless of whether the cooperative is active or passive.

The post-integration solution for the passive cooperative, represented by point 6, yields the most beneficial results to consumers. Output is the greatest and the final product price is the lowest of all solutions. Although the net price paid producers is greater than for either of the pre-integration solutions, producers would be better off with an active cooperative. The post-integration solution for the active cooperative maximizes joint profits π_{ABC}. Whereas the integrated passive cooperative behaves like a competitive firm in the final product market, the integrated active cooperative is a monopolist As a result, it receives a higher price for its final product.

Quantitative Solutions

Although the graphical analysis based on the fixed-proportions model has been useful in comparing the prices and output levels determined by the solution conditions in Table 3.2, we now return to the variable-proportions model and a quantitative analysis based on specific functional forms. By assuming specific functions and parameters, we can construct a more concrete illustration of the solutions through the generation of numerical values for prices, outputs, and other variables. We begin by assuming that the processor's production function is of the Cobb-Douglas form:

$$q_C = A q_B^\alpha w^{(1-\alpha)} \qquad A>0;\ 0<\alpha<1 \qquad (34)$$

and that it faces a final product demand curve of the form:

$$q_C = Z p_C^{-\epsilon} \qquad Z>0;\ \epsilon>1 \qquad (35)$$

where ϵ is the elasticity of demand. Consumer surplus is determined by calculating the area under the demand curve and subtracting $p_C q_C$, which is the consumer outlay for the final product:

$$CS = \int_0^{q_C} p_C \, dq_C - p_C q_C = \int_0^{q_C} Z^{\frac{1}{\epsilon}} q_C^{-\frac{1}{\epsilon}} dq_C - p_C q_C$$

$$= \left(\frac{\epsilon}{\epsilon-1}\right) Z^{\frac{1}{\epsilon}} q_C^{\left(1-\frac{1}{\epsilon}\right)} - p_C q_C \qquad (36)$$

We assume the assembler faces a linear supply curve. To construct the supply curve, we assume that total on-farm production costs take this form:

$$F = \int_0^{q_A} \left(\frac{1}{f} q_A - \frac{e}{f}\right) dq_A = \frac{1}{2f} q_A^2 - \frac{e}{f} q_A + g \qquad e \leq 0;\ f,\ g>0 \qquad (37)$$

where the constant of integration g represents fixed costs. The producer maximizes profit by setting the marginal cost of production equal to the price offered by the assembler:

$$MC_A = \frac{dF}{dq_A} = \frac{1}{f}q_A - \frac{e}{f} = p_A.$$ (38)

For convenience, in the case of a passive cooperative, we set the price p_A such that the per-unit patronage refund s is zero, which is consistent with pricing in the Helmberger and Hoos model. Solving (38) for q_A, the supply function facing the assembler is

$$q_A = e + fp_A.$$ (39)

Solutions are presented in Table 3.3 for the parameters shown at the foot of the table, which were chosen for illustrative purposes. Observations about prices and output levels based on the table generally conform to those drawn from the graphical analysis. In fact, some generalizations about integration and the three types of assemblers can be derived from experiments based on the quantitative model and the preceding graphical analysis.

Both consumers and producers are better off if the assembler is an active cooperative rather than an IOF. Whereas an IOF assembler is a monopsonist in the raw product market, an active cooperative assembler behaves like a competitive firm. Thus output and raw product price are greater and final product price is less. Because any profits of the cooperative assembler are returned to producers as patronage refunds, the net price received by producers may be further enhanced.

Consumers are always better off or just as well off if the assembler is a passive cooperative instead of an active cooperative. Whereas an active cooperative assembler behaves like a monopolist in the assembled and final product markets, a passive cooperative assembler behaves like a competitive firm. Thus output is greater and the final product price is less, except under processor dominance, for which output and price are the same.

A similar generalization cannot be made about the effect on producers. Although producer output is always at least as great for a passive cooperative, the receipt of patronage refunds complicates the comparison of producer revenues. Under assembler dominance, the output associated with the passive cooperative assembler is greater than that for the active cooperative. However, the net price paid producers by the active cooperative assembler is greater than that paid by the passive cooperative. Whether producers will be better off with an active or passive cooperative will depend on the specific demand and cost functions. Under processor dominance, the solutions for active and passive cooperatives are identical, and the net price paid producers is the same. If the

TABLE 3.3 Quantities, Prices, and Profits Under Alternative Market Structures and Variable-Proportions Technology

	Investor-Owned Firm			Active Cooperative			Passive Cooperative		
	Assembler Dominance	Processor Dominance	Post-Integration	Assembler Dominance	Processor Dominance	Post-Integration	Assembler Dominance	Processor Dominance	Post-Integration
	Million								
q_A, q_B	1.89	2.45	3.57	2.40	3.57	4.98	4.98	3.57	8.92
q_C	1.49	1.91	2.70	1.87	2.70	3.68	3.68	2.70	6.34
w	0.74	0.89	1.17	0.88	1.17	1.49	1.49	1.17	2.28
	Dollars								
p_A	0.68	0.70	0.74	0.70	0.74	0.80	0.80	0.74	0.96
s				0.30	0.00	0.34	0.00	0.00	0.00
p_A+s				1.00	0.74	1.14	0.80	0.74	0.96
p_B	1.17	0.80		1.10	0.84		0.90	0.84	
p_C	2.55	2.41	2.23	2.42	2.23	2.08	2.08	2.23	1.85
				Million Dollars					
π_A	0.07	0.12	0.25	0.12	0.25	0.50	0.50	0.25	1.59
π_B	0.75	0.00		0.73	0.00		0.00	0.00	
π_C	0.84	1.75		1.01	1.85		1.70	1.85	
π_{AB}				0.84	0.25		0.50	0.25	
π_{BC}			1.85			1.70			0.00
π_{ABC} Consumer						2.20			1.59
Surplus	1.08	1.31	1.72	1.29	1.72	2.19	2.19	1.72	3.34
Welfare	2.75	3.18	3.82	3.14	3.82	4.39	4.39	3.82	4.93

Parameters: $A=1$, $e=-15$, $f=25$, $h=0.1$, $r=1$, $Z=100$, $\alpha=0.75$, $\epsilon=4.5$.

cooperative integrates, producers are better off if the cooperative is active because this solution is associated with maximum joint profits π_{ABC}.

Both consumers and producers are better off after integration regardless of the assembler type. For each assembler, the post-integration solution yields a greater output and raw product price and a lower final product price than either pre-integration solution. Again the receipt of patronage refunds complicates the comparisons for cooperative assemblers. However, if the assembler is an active cooperative, producers will be better off after integration because the post-integration solution is associated with maximum joint profits π_{ABC}. If the assembler is a passive cooperative, the net price paid producers is greatest for the post-integration solution.

In addition to making both consumers and producers better off, integration increases total economic welfare. Integration by an IOF assembler results in the maximization of π_{BC}, the joint profits from assembling and processing the raw product. Thus, given that both consumers and producers are better off after integration, total welfare is increased. Integration by an active cooperative results in maximization of π_{ABC}, the joint profits from producing, assembling, and processing the raw product. Consequently, given that consumers are better off after integration, total welfare is once again increased. Economic welfare is the greatest when a passive cooperative integrates because the cooperative acts like a competitive firm in both the raw and final product markets. Welfare after integration is less when the assembler is an active cooperative because, although it behaves like a competitive firm in the raw product market, it exercises monopoly power in the final product market. Post-integration welfare is lowest when the assembler is an IOF because the assembler is a monopolist in the final product market and a monopsonist in the raw product market.

The vertical market structure preferred from a societal perspective would result from integration by a passive cooperative assembler because total economic welfare is greatest. Producers would prefer integration by an active cooperative because joint profits π_{ABC} are greatest under that structure. Both consumers and producers would prefer either structure to integration by an IOF assembler. Whether vertical integration is likely to arise at all will depend not on the desirability of the outcome but the incentives for the assembler to integrate, which will be explored in the following section.

Incentives to Integrate

We consider an assembler to have an incentive to integrate forward by acquiring the processor if the capitalized value of its objective function after integration, less what it must pay the owners of the processing plant, is greater than the capitalized value of its objective function before integration. The price the assembler must pay the owners of the processing plant will depend on the relative bargaining power of the two parties. However, under most circumstances, the minimum the owners of the plant would be willing to accept is the capitalized value of the plant's profits. Thus, if for simplification we assume that current profits are proportional to the capitalized values, an IOF has an incentive to integrate forward only if

$$\pi_{BC}^{*} - \pi_C > \pi_B \tag{40}$$

where π^* represents a post-integration profit.[2] Under the same conditions, a cooperative has an incentive to integrate only if

$$\pi_A^{*} + \pi_{BC}^{*} - \pi_C > \pi_A + \pi_B. \tag{41}$$

In the case of the IOF assembler, the assembler may have an incentive to integrate forward under both assembler and processor dominance. Because the integrated firm would maximize π_{BC}, condition (40) would be satisfied. In a similar manner, the active cooperative assembler may have an incentive to integrate forward because the integrated cooperative would maximize joint profits π_{ABC}, thereby satisfying condition (41). This is illustrated in Table 3.3. Under assembler dominance, $\pi_{BC}^{*} > \pi_C$ and the cooperative could finance acquisition of the processing plant by reducing the patronage refund shown in the table. Under processor dominance, $\pi_{BC}^{*} < \pi_C$ and the cooperative would have to finance the acquisition by eliminating the patronage refund and reducing the cash price paid for the raw product. Nevertheless, producers would be better off after the acquisition.

The outcome would likely be different for a passive cooperative assembler. In Tables 3.1 and 3.2, the solution conditions for the passive cooperative under assembler dominance are identical to the post-integration conditions for the active cooperative, which correspond to maximization of joint profits π_{ABC}. Thus, under assembler dominance, the passive cooperative would not have an incentive to integrate because integration would reduce π_{ABC} and condition (41) would not be satisfied. In general, we do not know whether condition (41) would be satisfied for a passive cooperative under processor dominance. Consequently, we cannot determine whether the cooperative would have an incentive to integrate without knowing the specific demand and cost functions. Condition (41) is not satisfied in the example in Table 3.3. Thus, unless the cooperative could restrict the quantity of raw product it handles through use of a nonprice mechanism, such as a delivery quota, processing right, or penalty scheme (Lopez and Spreen 1985, p. 389), it would not have an incentive to integrate forward into processing.

Conclusions

Both producers and consumers are better off when the assembler in a bilateral monopoly market structure is a cooperative. An active cooperative assembler results in a greater raw product price and output level and a lower final product price than an IOF assembler regardless of whether the assembled raw product market is dominated by the assembler or the processor. Although producers are usually better off when the assembler is an active cooperative than when it is a passive cooperative, overall economic welfare is maximized when the assembler is a passive cooperative.

These results are based on the differences in behavior of the three organizational forms. Given a monopsony situation in the raw product market and a monopoly situation in the assembled product or final product market, an IOF would be expected to exercise market power in both markets. Under identical circumstances, an active cooperative would be expected to behave like a competitive firm in the raw product market and as a monopolist in the assembled or final product market. A passive cooperative would behave like a competitive firm in both markets.

Given the benefits produced by cooperatives in bilateral monopoly markets, an argument could be made for public policy support of the existence and formation of cooperatives as well as the forward integration of cooperatives into processing activities. This would include support for active cooperatives, even if they exercise monopoly power in the final product market.

The critical determinant of the incentive for cooperatives to integrate forward into processing activities is whether they are successful in restricting producer output to optimal levels. If they are, integrating forward into the consumer market allows them to capture monopoly profits. Following the classic Helmberger and Hoos model of a marketing cooperative, cooperative theorists have argued that the receipt of patronage refunds provides producers an incentive to expand output until the cooperative breaks even. Under these circumstances, the cooperative will behave like a competitive firm. Unless it can restrict the quantity of raw product it handles using a nonprice mechanism, it will have no incentive to integrate forward into processed markets. This result provides an additional explanation, based on market power for the relatively low degree of cooperative forward integration.

Notes

1. Note that $MFC_A = d(MC_A q_A)/dq_A$ in the case of a cooperative assembler.
2. An exception to this and the following rule would exist if the assembler could construct a new processing plant for less than the capitalized value of the existing plant's profits. Assuming that the assembler would be successful in redirecting the entire raw

pıoduct supply to the new plant, recognition of this threat would force the owners of the existing plant to consider its replacement cost as the minimum they would accept.

References

Bhuyan, S., and J. S. Royer. 1992. Cooperative Incentives for Vertical Integration: The Bilateral Monopoly Case. Paper presented at the annual meeting of the American Agricultural Economics Association, Baltimore, Md., Aug. 9-12.

Caves, R. E., and B. C. Petersen. 1986. Cooperatives' Shares in Farm Industries: Organizational and Policy Factors. *Agribusiness: An International Journal.* 2(1):1-19.

Cotterill, R. W. 1987. Agricultural Cooperatives: A Unified Theory of Pricing, Finance, and Investment. In *Cooperative Theory: New Approaches*, ed. J. S. Royer, 171-258. Washington, D.C.: USDA ACS Serv. Rep. 18.

Eisenstat, P., and R. T. Masson. 1978. Capper-Volstead and Milk Cooperative Market Power: Some Theoretical Issues. In *Agricultural Cooperatives and the Public Interest*, ed. B. W. Marion, 51-68. N. Cent. Reg. Res. Pub. 256, University of Wisconsin-Madison.

Helmberger, P., and S. Hoos. 1962. Cooperative Enterprise and Organization Theory. *Journal of Farm Economics.* 44(2):275-90.

Jamison, J. A. 1960. Coordination and Vertical Expansion in Marketing Cooperatives. *Journal of Farm Economics.* 42(3):555-66.

Lopez, R. A., and T. H. Spreen. 1985. Co-ordination Strategies and Non-Members' Trade in Processing Co-operatives. *Journal of Agricultural Economics.* 36(3):385-96.

Perry, M. K. 1989. Vertical Integration: Determinants and Effects. In *Handbook of Industrial Organization,* vol. 1, ed. R. Schmalensee and R. D. Willig, 183-255. Amsterdam: North-Holland.

Rogers, R. T., and B. W. Marion. 1990. Food Manufacturing Activities of the Largest Agricultural Cooperatives: Market Power and Strategic Behavior Implications. *Journal of Agricultural Cooperation.* 5:59-73.

Schmiesing, B. H. 1989. Theory of Marketing Cooperatives and Decision Making. In *Cooperatives in Agriculture*, ed. D. W. Cobia, 156-73. Englewood Cliffs, N.J.: Prentice-Hall.

Sheldon, I. M. 1991. Vertical Coordination: An Overview. Paper presented at NC-194 symposium, Chicago, Ill., Oct. 17-18.

Staatz, J. M. 1989. *Farmer Cooperative Theory: Recent Developments.* Washington, D.C.: USDA ACS Res. Rep. 84.

4

Advertising Strategies by Agricultural Cooperatives in Branded Food Products, 1967 and 1987

Richard T. Rogers

Advertising is a major competitive strategy among leading food processors with the food processing sector outspending every other sector of the economy (Connor *et al.*, 1985). Advertising is the most successful method of creating and maintaining product differentiation within the food system. In fact, the advertising-to-sales ratio is often used as a proxy measure for the degree of product differentiation found in a processed food industry.

Firms choose advertising strategies along with pricing and product strategies in accordance with the market's basic conditions (e.g., consumer good) and structure (e.g., concentration). In tight oligopolies selling consumer nondurables to households, advertising is often the primary competitive weapon for the leading firms. Unlike price competition that leaves all rivals in the leading strategic group worse off, intense advertising rivalry by the leading firms can actually benefit themselves collectively. The advertising rivalry enhances the barriers to nonleading firms who operate in the market in a different strategic group. For example, Coke and Pepsi can benefit each other through their intense advertising rivalry as they make it more difficult for lesser known brands and makers of private label cola to enter the top strategic group of national branded colas. The former Pepsi president, John Scully, when asked about the "Cola Wars" said that such marketing battles do not involve "some gladiatorial contest where one of us has to leave on a stretcher. We're both winning." (*Wall Street Journal*, November 6, 1982, p. 1)

These advertising strategies can influence the market's future structure—product differentiation, concentration, and barriers to entry. In addition, economists include the extent of advertising in evaluating a market's performance. Although economists have not reached a concensus when advertising levels become excessive, Brandow offered the benchmark of 3 percent of sales as the point where advertising expenses become excessive. We will return to this benchmark when we compare the advertising expenditures of agricultural cooperatives to other food and tobacco processors.

Little is known about the advertising strategies used by the largest agricultural cooperatives that market branded food products. A growing literature on generic advertising for farm commodities, like milk and California prunes, examines industry issues related to advertising, but does not address branded advertising by agricultural cooperatives. Since the cooperatives are farmer-owned, there are many similarities between farmers voting for their industry-wide association (e.g., the American Dairy Association) to spend more on an industry-wide advertising campaign and supporting their cooperative in advertising the cooperative's brand (e.g., Land O'Lakes butter). Indeed, when an agricultural cooperative has a leading consumer brand (e.g., SunMaid in raisins) the cooperative's membership can choose an advertising strategy specific to its brand rather than, or in addition to, an industry-wide advertising campaign (e.g., California's dancing raisins).

An earlier study by Boynton shed some light on the extent of advertising used by agricultural cooperatives that market branded food products to household consumers. His study used 1979 data that included six measured media from the same advertising data source used here. He found 39 agricultural cooperatives used media advertising and that, in general, cooperatives spent substantially less than noncooperatives, even in industries where they were direct competitors.

The objective of this paper is to expand on Boynton's work and examine the advertising of branded products by agricultural cooperatives in food processing over a 20-year period, from 1967 to 1987. The related issue of generic, industry-wide advertising by associations or boards receives less attention here. Such advertising seeks to expand industry demand for the commodity as opposed to influence a consumer's brand choice among the various sellers. The 1987 data, unlike the 1967 data, allow some observations to be drawn about nonbrand advertising done on behalf of an entire industry, but the primary focus is on brand-specific advertising that attempts to build and maintain product differentiation.

The Food and Tobacco Processing Sector

The food processing sector dominates the other vertical stages of the food marketing system. More than 80 percent of domestically produced farm products flow through this sector as it transforms the raw agricultural products into processed products. The major farm products sold to household consumers in conventional supermarkets that do not pass through this sector are the fresh

TABLE 4.1 An Example of the Standard Industrial Classification (SIC) System

Level of Detail	Description	SIC #	Name
Two digit:	Major Group	20	Food and Kindred Products
Three digit:	Minor Group	202	Dairy Products
Four digit:	Industry	2026	Fluid Milk
Five digit:	Product Class	20262	Packaged Fluid Milk and Related Products
Seven digit:	Product	2026245	Sour Cream, Unflavored

Source: U.S. Department of Commerce, Bureau of the Census.

fruits, vegetables, eggs, and nuts. The government's Standard Industrial Classification System, the familiar SIC system, is used in this study to define the food and tobacco processing sector and the industries that comprise it. In this paper we often combine food processing, the SIC major group 20, with tobacco processing, SIC 21, unless noted otherwise. The SIC system assigns products a code number whose first two digits identify the products' major group and, as digits are added, the product is more narrowly classified. For example, in Table 4.1 the two-digit major group SIC 20 represents all processed food products. All dairy products belong in the minor group SIC 202, whereas all fluid milk, both consumer packages and bulk shipments belong in the four-digit industry SIC 2026. If the milk was packaged for household consumers, it would be in the five-digit product class SIC 20262. Finally, if the dairy product was unflavored sour cream, then it would receive the seven-digit product code, the narrowest classification used by the SIC system, SIC 2026245. The SIC system is extremely useful and forms the basis of government reports on economic activity by area. The system has been copied by many countries and private data vendors also use it. The system allows for easy aggregation from the product level to broader groupings of related products.

However, the SIC system is not perfect and economic researchers must learn some of its weaknesses. Economic markets are often best described by the five-digit product class level of detail, but there are many exceptions. For some markets the four-digit industry level is more appropriate (e.g., beer) while others require greater detail found with the seven-digit product (e.g., honey) for the best correspondence to an economic market. In some cases SICs must be combined to create a new category to align with an economic market (e.g., refined sugar requires combining the beet and cane sugar refining industries). Unfortunately, the amount of data that exists declines as the level of detail increases from the four-digit industry level to the seven-digit product level. The reason for discussing this data classification system is that cooperatives often operate in markets that require the finer detail of the seven-digit product level to reveal the cooperative's significance to the market (e.g., honey).

The 1987 advertising data for this research are from the Leading National

Advertisers, Inc. (LNA) and the 1967 data are primarily from LNA, but researchers at the Federal Trade Commission supplemented the LNA data with other sources to expand its coverage from six to eight media (see Rogers, 1982 for details on the 1967 data). LNA uses its own product classification scheme and it differs from the SIC system in what it reports as food products. The LNA system more closely resembles the food products found in a typical supermarket and combines processed food products with fresh unprocessed products. In its food classification system it includes fresh fruits, vegetables, eggs, and nuts but excludes animal feeds, whereas the SIC system does the opposite. The SIC system classifies the fresh farm products as unprocessed agricultural products if only cleaning, grading and bagging are involved. The SIC system classifies such products outside the food processing major group and in the SIC 01 or 02 major groups. These differences do not involve much brand advertising except in the case of pet foods, but the difference is significant because agricultural cooperatives are commonly involved in these fresh food products. For example, fresh oranges are classified by the SIC system in SIC 0174, and refrigerated orange juice in SIC 2033, and frozen orange juice in SIC 2037. Only the last two are within the definition of food processing (SIC 20). The Sunkist cooperative is a market leader in only the fresh market and almost all of its advertising is spent in support of its fresh oranges. Under the LNA classification scheme Sunkist's advertising for its fresh oranges would be combined with processed food products.

Another example will show the importance, but also the difficulty, a researcher faces in classifying the advertising expenditures for some food products that have a fine line between being classified as unprocessed food products or as processed food. Advertising for almonds or walnuts is classified in SIC 0173 if the nuts are only cleaned, shelled and bagged. But they are classified in SIC 2068, salted and roasted nuts and seeds, if they have been roasted or otherwise processed. This example is critical to the Blue Diamond and Sun-Diamond cooperatives.

There are some other minor differences between a LNA approach to classifying food and tobacco products and using a SIC approach, but unlike the above examples these do not have a major impact on research involving agricultural cooperatives. Bottled water is classified in food (beverages) by LNA, but in the SIC system it is in SIC 5149 (wholesaling) unless it is processed in some way. The processor decides if it is selling processed water and many firms selling bottled water prefer not to be considered as using processing as they wish to tout the natural qualities of their water. Another difference involves how cough drops are classified. They are in SIC 20649 or in SIC 28344 depending on the amount of sugar they contain. LNA does not classify any cough drops in its food categories. Lastly, pepper is in SIC

TABLE 4.2 Food and Tobacco Media Advertising Totals, 1987

	$ Millions
LNA[a]	5,950.5
SIC 20,21[b]	5,814.5
SIC 01,02,20,21,5149[c]	5,864.0

[a]A broad definition using all LNA codes related to food and tobacco products, but the LNA food codes include some products that belong in SIC 28 (Chemicals).
[b]Census Major Groups for Food and Tobacco Processing.
[c]Adds Census Major Groups for Agricultural Production of Crops (SIC 01) and Livestock (SIC 02), plus the wholesaling of bottled water (SIC 5149).
Source: Leading National Advertisers, Inc., 1987.

2099E31, but salt is outside food processing and found in SIC 28991, but both are in LNA's food classification.

In this paper, mainly the SIC system is used. In 1987 data are available for the fresh food products found in LNA's food category but classified outside of food processing (SIC 20). The broadest definition of food and tobacco advertising used here includes all advertising for what LNA classifies as food (those LNA codes beginning with F), plus advertising for pet and animal feeds, and tobacco products. In 1987 this amounted to $5,950.5 million spent on the seven measured media advertising tracked by LNA (Table 4.2). If a SIC definition of food and tobacco processing is used (SIC 20 and 21), the amount is $5,814.5 million or 97.7 percent of the amount found with the most inclusive LNA approach. If one adds the advertising for fresh farm products (SIC 01 and 02) and bottled water (SIC 5149) to the food and tobacco advertising, the amount increases to $5,864.0 million or 98.5 percent of the most inclusive definition. The remaining difference is explained by LNA including some products in its food classification that the SIC system places in chemicals (e.g., artificial sweeteners).

Agricultural Cooperatives in Food and Tobacco Processing

The food and tobacco processing sector has been dominated by its largest firms during most of this century, but the degree of domination by the very largest firms has accelerated during the last 15 years. Although, the Census counts over 15,000 food and tobacco processing firms, the 100 largest have accounted for the bulk of the sector's economic activity and their dominance has increased over time. By 1987, the latest year data are available, the 100 largest accounted for nearly 70 percent of the sector's value-added (Figure 4.1). Even among the 100 largest, it's the largest of the large that account for this increased

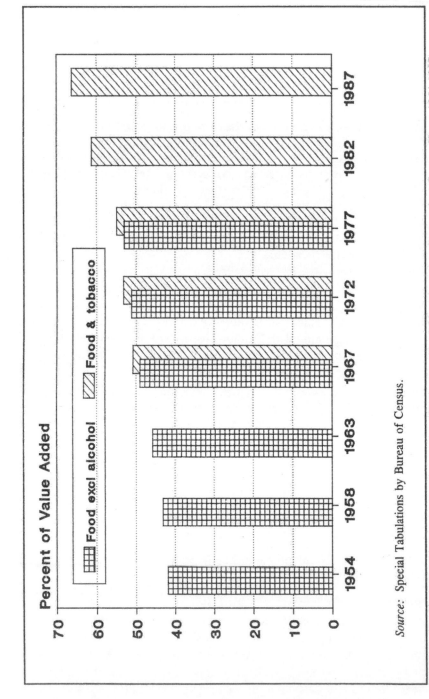

FIGURE 4.1 Aggregrate Concentration Among the 100 Largest Food Manufacturing Companies, Census Years 1954-1987

Source: Special Tabulations by Bureau of Census.

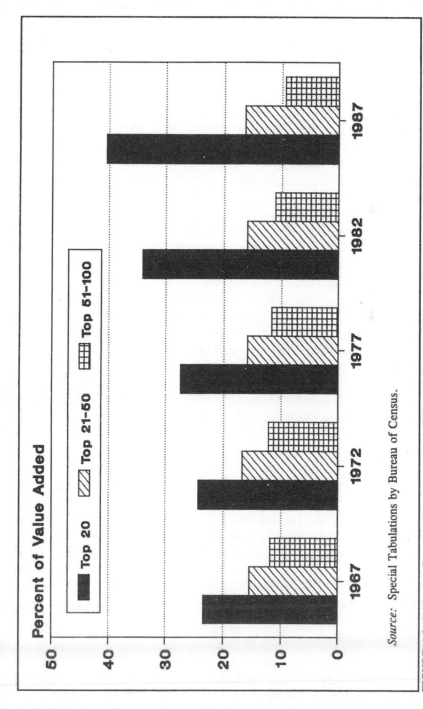

FIGURE 4.2 Aggregate Concentration Among the 100 Largest Food and Tobacco Manufacturing Companies, Census Years 1967-1987

domination. The 20 largest food and tobacco firms increased their share of the sector's value-added to over 40 percent by 1987, whereas firms ranked 21 to 50 largest and those ranked 51 to 100 barely held their own over the twenty-year period from 1967 to 1987 (Figure 4.2).

Agricultural cooperatives have also grown in size during this period and some have even called for a reexamination of public policy regarding cooperatives because of this growth (see Rogers and Marion, 1990). However, within the food and tobacco processing sector agricultural cooperatives have not kept pace with the largest food processing firms. Rogers and Marion found that there were no agricultural cooperatives among the 50 largest food and tobacco processors in 1982 measured by value-added. Only 4 of the 100 largest agricultural cooperatives were among those food and tobacco firms ranked 51 to 100 on the basis of value-added. Since cooperatives are often more prevalent in the commodity-oriented markets, they ranked higher when sales rather than value-added was the size measure, but no cooperatives were among the twenty largest food processors in 1982 based on food sales and only four ranked in the 21 to 50 largest group (Rogers and Marion).

Nevertheless, agricultural cooperatives have a significant presence in food processing. In 1982, 68 of the 100 largest agricultural cooperatives were involved in food processing and accounted for 7.2 percent of the sector's value of shipments (Table 4.3). Their combined share of shipments was higher in the more commodity-oriented products that involved minimal processing and used large volumes of their members' output. For example, the 100 largest cooperatives held 53 percent of the butter industry yet held none of the highly differentiated breakfast cereal industry.

The 20 largest food and tobacco firms prefer the more differentiated product markets. In 1982 the 20 largest firms, which do not include any agricultural cooperatives, held less than 10 percent of the value-added in industries classified as having none or low product differentiation, but their share of value-added was nearly 24 percent for the industries with medium product differentiation and nearly 47 percent for the 16 most highly differentiated industries (Table 4.4). The pattern for agricultural cooperatives was reversed. The 100 largest cooperatives held their greatest share of value-added in the undifferentiated industries and only held a slight 0.3 percent share of the value-added in the most highly differentiated industries.

Media Advertising by Agricultural Cooperatives

Although the commodity orientation of most agricultural cooperatives clearly shapes their use of advertising, other factors are involved. Theoretic economic models of advertising link firm advertising intensity to its price-cost margin, advertising elasticity, market and firm demand elasticities, and rivals' reactions.

TABLE 4.3 The 100 Largest Agricultural Cooperatives' Activity in Food and Tobacco Manufacturing at the 2-Digit, 3-Digit and Selected 4-Digit SIC Levels for 1977 and 1982

SIC	Name	# of Cooperatives[a]		Percentage of Universe Total Value of Shipments --percent--		Percent point change
		1982	1977	1982	1977	1982-1977
20	Food and Kindred Products	68	71	7.2	6.0	1.2
201	Meat Products	6	9	4.2	2.3	2.0
202	Dairy Products	32	28	24.4	17.7	6.7
2021	Butter	22	19	53.2	43.1	10.1
2022	Cheese, Natural and Processed	19	18	24.0	16.7	7.4
2023	Condensed and Evaporated Milk	31	25	34.0	27.3	6.7
2024	Ice Cream and Ices	16	18	7.7	5.2	2.5
2026	Fluid Milk	29	27	21.3	15.6	5.7
203	Preserved Fruits and Vegetables	27	32	8.9	8.3	0.6
2033	Canned Fruits and Vegetables	23	24	17.0	13.7	3.3
2037	Frozen Fruits and Vegetables	9	10	7.7	10.4	-2.7
204	Grain Mill Products	23	25	6.4	7.1	-0.7
2048	Prepared Feeds	18	18	10.7	12.0	-1.3
205	Bakery Products	0	0	0.0	0.0	0.0
206	Sugar and Confectionery Products	7	6	6.7	5.9	0.8
207	Fats and Oils	12	16	9.6	12.0	-2.4
2075	Soybean Oil Mill Products	8	8	15.2	N/A	N/A
208	Beverages	27	23	1.4	0.9	0.5
2086	Bottled and Canned Soft Drinks	21	21	1.9	N/A	N/A
209	Misc. Foods and Kindred Products	18	8	0.4	0.4	0.0
2099	Prepared Foods, N.E.C.	17	8	.8	N/A	N/A
21	Tobacco Products	0	0	0.0	0.0	0.0

Cooperatives are ranked by their value of sales in SIC 20, 21, 514 (except 5141) and 515. Five digit product class value of shipments data have been used in calculating percentages.
[a]Number of cooperatives from the top 100 sample processing some output in this industry group or industry. N/A = Not Available.
Source: Rogers and Torgerson, 1988.

TABLE 4.4 The Largest 100 Agricultural Marketing Cooperatives' Activity in National Food and Tobacco Industries by the Degree of Product Differentiation, 1982

Degree of Product Differentiation		Percent of Value-Added		
		Top 20 Investor-Owned Companies	Top 20 Cooperatives	Top 100 Cooperatives
None	(10)	5.3	3.7	8.0
Low	(6)	3.8	4.2	4.9
Medium	(13)	23.8	2.8	6.2
High	(16)	46.8	0.2	0.3

Note: The number in parentheses is the number of national industries that are classified in this product differentiation group. The six local industries not included here are: 2024, 2026, 2048, 2051, 2086, 2097. None of the 100 cooperatives was in 2051 or 2097.

Source: Rogers and Torgerson, 1988.

The premier theoretical result was derived nearly 40 years ago by Dorfman and Steiner where they showed that a monopolist's optimal advertising-to-sales ratio was equal to the ratio of the price elasticity of demand to the advertising elasticity of demand. This can be rewritten as the price-cost margin times the advertising elasticity of demand. This result has been extended to the oligopoly case where one has to add the effect of rivals but the main result remains that advertising will be higher, the larger the industry's sales, the higher the profit margin, the higher the sales response to advertising, and the more price inelastic the product is.

To date there is nothing in the theory that suggests an agricultural cooperative should behave differently than any other firm. The commodity orientation common to most cooperatives should be captured by the same industry factors, such as the price-cost margin and the product's response to advertising, that an investor-owned firm may face. Yet it is hypothesized that there is more to it than is specified in the theory. For all else equal, the average cooperative is hypothesized to spend less on advertising than an investor-owned firm. Reasons for this may include less emphasis on marketing than is often found in investor-owned firms and a greater difficulty in getting managerial, board and general membership approval for such expenditures.

The LNA data are restricted to the main measured media targeted at wide consumer audiences. The seven media covered in 1987 were network television, spot television, cable television, network radio, magazines, outdoor, and Sunday newspaper supplements (e.g., Parade). Only the network advertising is continuously monitored, most of the media are represented only by selected markets or leading publications. Using time and space measurements of the advertisements, the advertising expenditures are estimated and assigned to company and brand records. Only those companies, or brands if the parent company cannot be identified, that spent at least $25,000 in the year are included in their publications (see Rogers, 1982 for a full description of the LNA data). Researchers have found that the measured media totals represent anywhere from 30 to 50 percent of the total selling expenses involved in promoting food products (Connor *et al.*, 1985). Although substantial amounts of promotional expenses are omitted, the measured media data offer the best available data that are clearly directed at the creation and maintenance of product differentiation. In addition, research has shown it to be positively correlated with other selling expenses (again, see Connor *et al.*, 1985).

As already discussed, the broadest definition of food and tobacco products gave the 1987 total expenditure on these media as $5,950.5 million. Agricultural cooperatives accounted for $73.6 million, or 1.238 percent of the total (Table 4.5). Advertising by associations include all nonbrand specific advertising done by industry associations, states or countries, and boards or commissions on behalf of an entire industry rather than a particular firm. The advertising by associations far exceeded that done by cooperatives and amounted to $215.8 million, or 3.627 percent of the total food and tobacco media advertising. Since the focus of this paper is on brand advertising, the last column of Table 4.7 gives the percent of the total advertising held by cooperatives after netting out all advertising done by the associations. This percentage reflects the share of branded advertising done by cooperatives in food and tobacco industries.

The shares held by cooperatives or associations changed little by restricting the definition of food processing to the Census definition (SIC 20) and neither associations nor cooperatives advertised tobacco products (SIC 21). The largest percentage share held by cooperatives was in fresh produce (SIC 01), with about a 25 percent share of the total $33 million, which included $20 million of association advertising. The cooperative share jumped to nearly 65 percent if such nonbrand advertising was omitted. The total amount spent in these unprocessed industries, however, is still quite small in comparison to the amounts found in processed food industries. In addition, one cooperative, Sunkist, accounted for nearly half of the $8.4 million spent in SIC 01. Even less was spent supporting products classified in SIC 02, livestock products, and almost all (96.5 percent) of the nearly $3 million spent was done by associations.

Although agricultural cooperatives accounted for 1.3 percent of the total advertising for branded food products, their share within the nine industry minor groups that comprise food processing (SIC 20) varied from nearly zero in the grain mill and bakery products (SICs 204 and 205) to over five percent in the preserved fruits and vegetables minor group (again see Table 4.5). Cooperatives spent $37.3 million, or over half (57 percent) of their total food processing advertising in this minor group (SIC 203) but their share of the branded

TABLE 4.5 Media Advertising Totals for Food and Tobacco Products, 1987

SIC	Description	All Advertisers ($ millions)	Agricultural Cooperatives		Associations		Cooperative's Share Net of Associations (percent)
			$ millions	percent	$ millions	percent	
Total [a]	All Food and Tobacco Related Advertising	5,950.5	73.6	1.238	215.8	3.627	1.284
20	Food and Kindred Products	5,286.2	65.2	1.234	192.8	3.648	1.281
21	Tobacco Manufacturers	528.3	0.0	0.000	0.0	0.000	0.000
20+21	Food and Tobacco Products	5,814.5	65.2	1.122	192.8	3.316	1.281
01	Agricultural Production—Crops	33.1	8.4	25.240	20.2	61.055	64.808
02	Agricultural Production—Livestock	2.9	0.1	2.130	2.8	96.546	61.662
01+02	Crops and Livestock	36.0	8.4	23.383	23.0	63.907	64.784
201	Meat Products	213.4	1.9	0.888	29.8	13.953	1.033
202	Dairy Products	347.6	6.5	1.871	114.7	33.005	2.792
203	Preserved Fruits and Vegetables	738.6	37.3	5.048	23.8	3.223	5.216
204	Grain Mill Products	947.9	1.0	0.106	0.0	0.003	0.106
205	Bakery Products	288.1	0.4	0.144	0.0	0.000	0.144
206	Sugar and Confectionery Products	487.9	6.6	1.345	1.7	0.350	1.350
207	Fats and Oils	93.5	2.4	2.546	0.0	0.000	2.546
208	Beverages	1,593.8	8.1	0.508	9.1	0.572	0.510
209	Misc. Foods and Kindred Products	563.4	0.8	0.135	13.4	2.380	0.138

[a] A broad definition using all LNA codes related to food and tobacco products, but the LNA food codes include some products that belong in SIC 28 (Chemicals).

Source: Leading National Advertisers, Inc., 1987.

advertising in this group was only 5.2 percent. This amount is sensitive to how one classifies Ocean Spray's advertising. The beverage products that Ocean Spray markets are difficult to classify between SIC 2033A or 2033B (canned or fresh juices) and SIC 20866 (canned and bottled drinks containing real juice). Nearly $16 million of the $37 million cooperatives spent here is from Ocean Spray's ready-to-serve juices. If they were classified in SIC 20866, the total spent by cooperatives would be roughly equal between the minor groups SIC 203 and SIC 208. The determining factor in classifying these drinks is the amount of real juice in the beverages. Only those fruit beverages with 100 percent real juice belong in SIC 2033A,B and all others belong in SIC 20866. Since cranberry juice must be diluted to be drinkable, most of Ocean Spray's traditional cranberry-based drinks belong in SIC 20886, but Ocean Spray now markets a wide array of 100 percent juice beverages that belong in SIC 2033A,B. This delicate classification distinction must be considered when examining cooperative advertising over time.

The next largest share held by cooperatives was in dairy products, SIC 202, where cooperatives spent $6.5 million, or 2.8 percent of branded dairy advertising. That amount was dramatically overshadowed by the advertising expenditures by associations which spent nearly $115 million, including nearly $94 million by the American Dairy Association. The cooperative share of branded advertising also exceeded 2 percent in fats and oils, SIC 207, with about $2.4 million spent. Land O'Lakes accounted for over $2 million, or 88 percent of this total, with its margarine advertising. Unlike dairy products, no association advertising was done in SIC 207.

In none of the nine industry minor groups did the cooperatives share of total advertising in 1987 exceed their share of sales in 1982 (except in SIC 205, bakery products, where none of the top 100 agricultural cooperatives operated in 1982, and in 1987 cooperatives held a 0.144 percent share of brand advertising because we assigned advertising for SunMaid cinnamon rolls and english muffins here even though the products are manufactured under a licensing agreement). For example, the 100 largest agricultural cooperatives held 24.4 percent of the value of shipments in dairy products but only a 2.8 percent share of branded advertising. The difference was closer in the other three-digit categories and it's quite close in SIC 203 and SIC 208, especially if one considers the delicate data classification question regarding fruit juices and juice drinks. This difference underscores the finding that cooperatives tend to leave the more differentiated consumer products to investor-owned firms even in industries where they operate processing plants (e.g., ice cream).

Associations spent over half of their total food related advertising in support of dairy products ($115 million), followed by meat products ($30 million), preserved fruits and vegetables ($24 million), and fresh produce ($20 million). These industry groups are comprised of commodity markets supplied by large numbers of farmers, many of whom belong to one or more cooperatives.

Whether these farmers are best served through advertising done by an industry-wide generic advertising campaign or by their cooperative advertising their individual branded product is a difficult research issue that is central to the federally-funded project NEC-63 on commodity advertising and promotion (see Kinnucan *et al.*, 1992 for more information). For commodity markets where there are virtually no brand names, like fresh beef, industry-wide campaigns are the most plausible way to advertise.

Comparing 1987 to 1967 Advertising

Unfortunately, some differences exist between the advertising data collected for 1987 and that available in 1967. In 1967 only processed food and tobacco (SICs 20 and 21) brand advertising was collected and hence we cannot make comparisons with 1987 advertising outside of food and tobacco processing. Also, no data was collected on nonbrand advertising done by associations, states, or boards, and, thus, no comparisons can be made with such data for 1987. The 1967 LNA data were originally compiled by the personnel at the Federal Trade Commission, where they supplemented media tracked by LNA by adding advertising data on spot radio and newspaper advertising beyond that already included in newspaper supplements. In 1987 no data were collected for spot radio and only newspaper supplements were included. By 1987 cable television had emerged as a new medium and LNA gathered information on advertising expenditures on this new media, but since cable television did not exist in 1967 comparisons are possible. Unfortunately, the 1967 data set combined newspaper advertising with advertising in newspaper supplements making any comparisons with 1987 newspaper supplements advertising inappropriate. The addition of spot radio was significant as it accounted for 6.8 percent of all branded food and tobacco media advertising whereas network radio accounted for only 1.1 percent and was the least used media in 1967 (Table 4.6).

Television advertising dominated the media included in both years, accounting for 65 percent of the total $1,611 million spent in 1967 on branded food and tobacco products (see Table 4.6). In 1987, television (spot, network, and cable combined) accounted for 75 percent of the $5,814.5 million spent on branded food and tobacco advertising. The increased share for television is partially explained by the 1987 data not including all newspaper advertising and spot radio advertising. The media with the largest percentage increase over the twenty-year period was network radio, but it grew from the smallest base in 1967. Outdoor advertising showed a similar increase. The next largest percentage increase was in network television, which just slightly exceeded spot TV as the largest media in 1967. By 1987 network television distanced itself from the other media as the preferred media for food and tobacco brand advertising. Cigarette advertising was allowed on television in 1967 but was

TABLE 4.6 Processed Food and Tobacco Advertising, by Media, 1967 and 1987

	1967		1987		Change (percent)
	Amount ($ millions)	Share (percent)	Amount ($ millions)	Share (percent)	
Network Radio	18.2	1.1	144.0	2.5	691.2
Spot Radio	109.5	6.8	N/A	N/A	N/A
Outdoor	39.5	2.5	256.7	4.4	549.9
Newspapers[a]	160.7	10.0	93.8	1.6	N/A
Magazines	230.4	14.3	964.9	16.6	318.8
Spot TV	523.4	32.5	1,573.0	27.1	200.5
Network TV	529.4	32.9	2,619.6	45.1	394.8
Cable TV	0.0	0.0	162.5	2.8	N/A
Total TV	1,052.8	65.3	4,355.1	74.9	313.7
TOTAL	1,611.1	100.0	5,814.5	100.0	260.9

[a] In 1967 local newspaper advertising was included with advertising in newspaper supplements by FTC, but in 1987 only newspaper supplements were included.
N/A = Not available or not appropriate comparison.
Source: Federal Trade Commission and Leading National Advertisers, Inc.

banned from television in 1971, so the increased use of television by other industries (except hard liquor that does not advertise on television by a self-imposed industry decision) was impressive. After television, magazines were the most heavily used for advertising branded food and tobacco products in both years. Television's dominance of the LNA media used for advertising branded food products is substantial. It is the best suited media for the creation and maintenance of product differentiation in branded food products and often firms tie their print media advertising to the same themes developed in their television advertising (see Connor *et al.*, 1985 for more information).

In 1967, 660 companies advertised branded food and tobacco products through the measured media tracked by LNA and 31 of these were agricultural cooperatives. By 1987, this had increased to nearly 800 companies, including 39 agricultural cooperatives. In 1967, these 31 cooperatives accounted for 1.31 percent of total branded media advertising in food processing (SIC 20). This share was essentially unchanged twenty years later as the 39 cooperatives held a 1.28 share of branded processed food advertising (Table 4.7). No cooperatives were involved in tobacco advertising in either year. Even when food processing is separated into its nine three-digit industry groups, there was little change in the shares held by cooperatives across this twenty-year period. What appears to be an increase in SIC 203 and a decrease in SIC 208 is more related to how Ocean Spray's products and advertising were allocated between juices and drinks as previously discussed. The share of branded advertising held by cooperatives in the dairy products group did increase from 1.6 percent to 2.8 percent. The increase in the fats and oils group, from nearly 0 to 2.5 percent is largely the result of Land O'Lakes advertising its margarine. The other comparisons have changes of less than half a percentage point but with most groups posting a minor share increase. As already discussed, the increased share in bakery products (SIC 205) was the result of a licensing arrangement to use the SunMaid name rather than direct bakery operations by an agricultural cooperative.

The Leading Advertisers in 1967 and 1987

Advertising expenditures are very concentrated among the leading advertisers in food and tobacco processing and the extent of concentration has increased substantially since 1967. The advertising concentration is much more dramatic than that found with value-added or sales. In 1967, the four largest advertisers accounted for 19.4 percent of all food and tobacco advertising and the 20 largest advertisers held a 53.4 percent share. By 1987 the top four's share had increased to 32.8 percent and the top 20 accounted for 72.1 percent (Table 4.8). The 50 largest advertisers in 1967 accounted for 78.1 percent, but by 1987 their share had increased to 90.6 percent. Although nearly 800 firms used media advertising in 1987, the top 100 advertisers accounted for 96.2 percent of the advertising expenditures.

TABLE 4.7 Percent of Total Media Advertising by Cooperatives in Broad SIC Categories, 1967 and 1987

SIC	Name	1967	1987	Percentage Point Change
		percent		
20	Food and Kindred Products	1.307	1.281	-0.026
21	Tobacco Manufacturers	0.000	0.000	0.000
20+21	Food and Tobacco Processing	1.063	1.160	0.097
201	Meat Products	0.623	1.033	0.410
202	Dairy Products	1.601	2.792	1.191
203	Preserved Fruits and Vegetables	3.451	5.216	1.765
204	Grain Mill Products	0.198	0.106	-0.092
205	Bakery Products	0.000	0.144	0.144
206	Sugar and Confectionery Products	0.929	1.350	0.421
207	Fats and Oils	0.050	2.546	2.496
208	Beverages	1.928	0.510	-1.427
209	Misc. Foods and Kindred Products	0.047	0.138	0.091

Source: Federal Trade Commission and Leading National Advertisers, Inc.

TABLE 4.8 Concentration of Media Advertising Expenditures in Food and Tobacco Processing, 1967 and 1987

Advertiser's Rank	1967 Share	1987 Share
	percent	
Top 4	19.4	32.8
Top 8	29.9	47.3
Top 20	53.4	72.1
Top 50	78.1	90.6
Top 100	90.5	96.2

Note: Excludes advertising by associations, boards, and governments.
Source: Leading National Advertisers, Inc.

The largest food and tobacco advertiser in 1967 was General Foods, now owned by Philip Morris Companies who was the largest advertiser in 1987. The 25 largest advertisers in 1967 are listed in Table 4.9. No agricultural cooperatives were among the fifty largest advertisers in 1967. The largest cooperative advertiser was Ocean Spray which was ranked 65th overall and

TABLE 4.9 Leading Company Advertisers in Food and Tobacco Processing, 1967

Rank	Company	Company Total ($000)	Percent of Total	Cumulative Percent
1	General Foods Corporation	105,408	6.543	6.543
2	Reynolds RJ Tobacco Company	77,979	4.840	11.383
3	American Tobacco Company	70,017	4.346	15.728
4	Coca-Cola Company	59,232	3.676	19.405
5	National Dairy Products Corporation	43,556	2.703	22.108
6	Kellogg Company	42,508	2.638	24.747
7	Lorillard P Company	41,469	2.574	27.321
8	Philip Morris Inc.	41,424	2.571	29.892
9	General Mills Inc.	40,455	2.511	32.403
10	Distillers Corporation-Seagrams Ltd.	39,197	2.433	34.836
11	Liggett & Myers Tobacco Company	38,061	2.362	37.198
12	Brown & Williamson Tobacco	37,994	2.358	39.556
13	Pepsico Inc.	37,235	2.311	41.867
14	Campbell Soup Company	31,794	1.973	43.841
15	Wrigley William Jr. Company	28,605	1.775	45.616
16	Quaker Oats Company	28,419	1.764	47.380
17	Standard Brands Inc.	26,672	1.655	49.036
18	Carnation Company	24,641	1.529	50.565
19	Lever Brothers Company	23,667	1.469	52.034
20	Procter & Gamble Company	22,208	1.378	53.412
21	Ralston Purina Company	21,659	1.344	54.757
22	Continental Baking Company Inc.	20,003	1.242	55.998
23	National Biscuit Company	19,870	1.233	57.232
24	Heublein Inc.	18,606	1.155	58.386
25	Corn Products Company	17,243	1.070	59.457

Source: Leading National Advertisers Inc.

accounted for 0.306 percent of total food and tobacco advertising (Table 4.10). Only four cooperatives placed among the 100 largest food and tobacco advertisers.

The leading company advertisers from 1967 were still among the leaders in 1987 after accounting for mergers and name changes (Table 4.11). In 1987 Philip Morris led all food and tobacco advertisers with over 13 percent of all food and tobacco advertising, more than double the share held by the largest advertiser in 1967. In 1988, Philip Morris acquired Kraft Inc., the tenth largest advertiser in 1987 with a 2.5 percent share of total food and tobacco advertising. Two companies, Coors and Hershey, were not among the top 25 advertisers in 1967 but by 1987 they had become major advertisers. Coors was often used as an example of a highly differentiated consumer product that had great success without media advertising. In 1967 Coors was only the 100th largest advertiser,

TABLE 4.10 Leading Agricultural Cooperative Advertisers in Food and Tobacco Processing, 1967

Rank	Company	Company Total ($000)	Percent of Total
65	Ocean Spray Cranberries Inc.	4,932	0.306
75	Allied Grape Growers	3,753	0.233
79	North Pacific Canners & Packers	3,587	0.223
84	National Grape Co-op Association Inc.	2,749	0.171
130	Sunsweet Growers Inc.	1,152	0.072
132	California Canners & Growers	1,128	0.070
143	Land O'Lakes Creameries Inc.	983	0.061
169	California & Hawaiian Sugar	634	0.039
172	Atlanta Dairies	619	0.038
208	Gold Kist	382	0.024
210	Arkansas Rice Growers Co-op Association	379	0.024

Source: Leading National Advertisers Inc.

but that changed in the 1970s as it began losing market share when Philip Morris and Anheuser-Busch escalated advertising rivalry in the beer industry. In 1987 it had become the 19th largest advertiser, a high ranking for such an undiversified company. Hershey increased its advertising even more so. Before the death of its founder, who opposed commercial advertising, Hershey was an extremely small advertiser given its size and market share in the candy market. In 1967 it ranked as only the 223rd largest advertiser but by 1987 it had risen to rank 24th and had embraced advertising as a competitive strategy.

The 1987 listing of the largest food and tobacco advertisers includes associations and other organizations that advertise generically an industry-wide message (e.g., Drink Milk). The largest such advertiser was the American Dairy Association which ranked 17th among all food and tobacco advertisers (Table 4.12). The National Livestock and Meat Board was a distant second, followed by the state of Florida. The top 25 such advertisers are listed in Table 4.12.

By 1987, 39 agricultural cooperatives advertised food products and the leading 25 ranked by food and tobacco advertising are given in Table 4.13. Just as was the case in 1967, Ocean Spray was the largest cooperative food and tobacco advertiser, and had risen in overall rank from 65th in 1967 to 42nd in 1987, if associations are omitted as they were in 1967. Whereas four agricultural cooperatives made the top 100 advertisers in 1967, six made the top 100 in 1987, as well as eight associations.

TABLE 4.11 Leading Company Advertisers in Food and Tobacco Processing (Including Associations), 1987

Rank	Company	Company Total ($000)	Percent of Total	Cumulative Percent
1	Philip Morris Companies Inc.	769,772.9	13.239	13.239
2	RJR Nabisco Inc.	439,313.9	7.556	20.794
3	Anheuser-Busch Cos Inc.	325,465.8	5.598	26.392
4	Kellogg Company	309,008.9	5.314	31.706
5	General Mills Inc.	263,425.0	4.530	36.237
6	Mars Inc.	203,544.5	3.501	39.738
7	Pepsico Inc.	174,923.2	3.008	42.746
8	Coca-Cola Company	171,662.2	2.952	45.698
9	Nestle Sa	164,886.9	2.836	48.534
10	Kraft Inc.	145,139.1	2.496	51.030
11	Procter & Gamble Company	136,576.6	2.349	53.379
12	Quaker Oats Company	133,101.6	2.289	55.668
13	Campbell Soup Company	129,571.9	2.228	57.897
14	Unilever NV	124,408.1	2.140	60.036
15	Ralston Purina Company	124,147.5	2.135	62.171
16	Wrigley WM Jr. Company	102,379.1	1.761	63.932
17	American Dairy Association	93,987.1	1.616	65.549
18	BCI Holdings Corporation	86,342.0	1.485	67.034
19	Coors Adolph Company	84,656.8	1.456	68.490
20	Heinz HJ Company	84,441.6	1.452	69.942
21	Grand Metropolitan PLC	78,014.5	1.342	71.284
22	Hicks & Haas	76,115.6	1.309	72.593
23	Gallo E & J Winery	63,322.3	1.089	73.682
24	Hershey Food Corporation	62,406.8	1.073	74.755
25	Seagram Company Ltd.	61,939.7	1.065	75.820

Source: Leading National Advertisers Inc., 1987.

One of the largest advertisers among the 39 agricultural cooperatives in 1987 was Sunkist who does not appear on the top 25 list in Table 4.13. This is because Sunkist allocated almost all its advertising to its fresh fruit, which is not classified in processed food (SIC 20). To overcome this narrower definition of food advertising, all 39 cooperatives that advertised food products in 1987 are listed in Table 4.14, ranked by their total food advertising, which includes the fresh products (SICs 01 and 02). Sunkist was the sixth largest advertiser with this broader definition. Ocean Spray remained the largest food advertiser even with the inclusion of unprocessed food products and accounted for 27.7 percent of the total food advertising expenditures by cooperatives. The inclusion of the fresh products that are not classified as processed food products only affects five cooperatives since the other 34 advertised only processed food products.

TABLE 4.12 Leading Association Advertisers in Food and Tobacco Processing, 1987

Rank	Association [a]	Company Total ($000)	Percent of Total
17	American Dairy Association	93,987.1	1.616
39	National Live Stock & Meat Board	29,074.3	0.500
53	Florida State of	13,657.8	0.235
56	National Federation of Coff Grwr of Columbia	11,211.4	0.193
62	National Dairy Promo & Research Board	9,054.8	0.156
64	Calif Oregon Wash Dairyman Association	8,714.1	0.150
75	Puerto Rico Commonwealth of	6,673.2	0.115
76	California Raisins Advisory Board	6,640.3	0.114
111	California Prunes Advisory Board	2,811.0	0.048
140	Italy Republic of	1,746.2	0.030
144	Sugar Association Inc.	1,706.9	0.029
156	Catfish Institute	1,393.9	0.024
189	California Milk Advisory Board	898.7	0.015
228	German Agricultural Marketing Board	568.4	0.010
251	Alaska Seafood Marketing Institute	402.9	0.007
254	Olive Administration Committee	397.3	0.007
265	American Sheep Producers Council Inc.	357.9	0.006
277	Switzerland Cheese Association	335.6	0.006
289	Calilfornia Dates Administrative Committee	300.7	0.005
317	Norwegian Sardine Indus	244.9	0.004
351	Milk for Health Agency Canada	181.4	0.003
356	United Dairy Association	168.1	0.003
393	Mid Atlantic Dairy Association	128.4	0.002
409	Vanilla Information Bureau	118.3	0.002
433	Norwegian Salmon Marketing Council	100.4	0.002

[a] An Association is used here to refer to any group advertising on behalf of an entire industry rather than company brands.
Source: Leading National Advertisers Inc., 1987.

Since television is considered the premium media for creation and maintenance of product differentiation, it is of interest whether cooperatives allocated most of their media advertising dollars to television. Of the leading 10 cooperative advertisers, four allocated at least 90 percent of their advertising to television. Interestingly, the fifth largest advertiser, Land O'Lakes, only allocated 14.5 percent of its advertising to television. Tri-Valley also made much less use of television than the average, spending only 23.5 percent on television. Ocean Spray, on the other hand, spent nearly all of its media advertising dollars on television (98 percent), as did California Almond Growers Exchange (now Blue Diamond) and Guild Winery.

Although the 1967 data did not include unprocessed food advertising, 31 cooperatives did advertise processed food products (Table 4.15). Ocean Spray

TABLE 4.13 Leading Agricultural Cooperative Advertisers in Food and Tobacco Processing, 1987

Rank	Company	Company Total ($000)	Percent of Total
44	Ocean Spray Cranberries Inc.	20,408.7	0.351
58	Sun-Diamond Growers of California	9,954.3	0.171
68	Agway Inc.	7,882.0	0.136
90	Land O'Lakes Inc.	4,627.8	0.080
96	Alexander & Baldwin Inc. (C & H Sugar)	3,677.6	0.063
99	Guild Winery & Distillers	3,502.0	0.060
109	California Almond Growers Exchange	2,885.3	0.050
113	National Grape Cooperative Association	2,641.6	0.045
149	Tri-Valley Growers	1,587.4	0.027
150	Tree Top Inc.	1,565.4	0.027
161	Farmland Inds Inc.	1,308.4	0.023
181	Citrus World Inc.	986.4	0.017
185	Gold Kist Inc.	911.2	0.016
268	Riceland Foods Inc.	353.0	0.006
276	Sioux Honey Association	336.7	0.006
278	United Dairymen/Arizona	332.8	0.006
283	Darigold Inc.	315.7	0.005
306	Coble Dairy	272.1	0.005
310	Challenge Dairy Products	263.2	0.005
347	Golden Guernsey Dairy Coop	183.5	0.003
375	Tillamook County Creamery Association	151.1	0.003
376	Cream O Weber Dairy Company	146.4	0.003
416	Prairie Farms Dairy Company	112.3	0.002
429	Dairymen Inc.	102.1	0.002
437	Upstate Milk Corporation Inc.	98.5	0.002

Source: Leading National Advertisers Inc., 1987.

was the largest cooperative advertiser and accounted for 24.2 percent of the total cooperative advertising. In 1967, seven cooperatives spent over a million dollars and the distribution of expenditures was less concentrated than in 1987. Of the 39 cooperatives that advertised in 1987, only twelve spent more than a million dollars (without adjusting for inflation). Sunkist and California Almond Growers Exchange (now Blue Diamond) were much lower on the 1967 list because of the omission of advertising for unprocessed food products. Also, in 1967 the Sun-Diamond marketing coalition had not been formed so SunMaid, Sunsweet, and Diamond Walnut were all listed separately.

Five of the top ten cooperative advertisers in 1967 allocated over 90 percent of their media advertising expenditures to television, but the tenth largest, Gold Kist, did not use any television advertising in 1967 whereas by 1987 television was the only media it used. In 1967, 13 of the 31 cooperatives did not use

TABLE 4.14 Media Food Advertising by Agricultural Cooperatives, 1987

Rank	Company	Total ($000)	%of Co-operatives Total	% SIC 20	% TV
1	Ocean Spray Cranberries Inc.	20,408.7	27.71	100.0	97.9
2	Sun-Diamond Growers of California	12,150.0	16.50	81.9	75.4
3	Agway Inc. (incl. HP Hood & C Burns)	7,882.0	10.70	100.0	62.5
4	California Almond (Blue Diamond)	5,110.6	6.94	43.5	100.0
5	Land O Lakes Inc.	4,627.8	6.28	100.0	14.5
6	Sunkist Growers Inc.	3,890.3	5.28	0.1	83.6
7	Alexander & Baldwin Inc. (C & H Sugar)	3,677.6	4.99	100.0	92.0
8	Guild Winery & Distillers	3,502.0	4.75	100.0	99.9
9	National Grape Cooperative Association	2,641.6	3.59	100.0	53.1
10	Tri-Valley Growers	1,587.4	2.16	100.0	23.5
11	Tree Top Inc.	1,565.4	2.13	100.0	94.0
12	Farmland Inds Inc.	1,308.4	1.78	100.0	99.7
13	Citrus World Inc.	986.4	1.34	100.0	100.0
14	Gold Kist Inc.	911.2	1.24	100.0	100.0
15	Riceland Foods Inc.	353.0	0.48	100.0	88.6
16	Sioux Honey Association	336.7	0.46	100.0	26.9
17	United Dairymen/Arizona	332.8	0.45	100.0	95.8
18	Darigold Inc.	315.7	0.43	100.0	100.0
19	Coble Dairy	272.1	0.37	100.0	100.0
20	Challenge Dairy Products	263.2	0.36	100.0	38.6
21	Golden Guernsey Dairy Coop	183.5	0.25	100.0	100.0
22	Tillamook County Creamery Association	151.1	0.21	100.0	27.7
23	Cream O Weber Dairy Company	146.4	0.20	100.0	100.0

(continues)

TABLE 4.14 *(continued)*

Rank	Company	Total ($000)	%of Co-operatives Total	% SIC 20	% TV
24	Prairie Farms Dairy Company	112.3	0.15	100.0	100.0
25	Dairymen Inc.	102.1	0.14	100.0	91.2
26	Upstate Milk Corporation Inc.	98.5	0.13	100.0	100.0
27	Cabot Farmers Co-Op Creamery Co. Inc.	79.1	0.11	100.0	94.2
28	Inner Mountain Egg Producers Association	61.6	0.08	0.0	100.0
29	Swiss Valley Farms	59.6	0.08	100.0	100.0
30	Norbest Turkey Growers Association	55.1	0.07	100.0	47.4
31	Sealed-Sweet Growers Inc.	49.7	0.07	0.0	100.0
32	Roberts Dairy Company	47.9	0.07	100.0	76.0
33	Bison Foods	47.8	0.06	100.0	100.0
34	Knouse Foods Inc.	41.0	0.06	100.0	0.0
35	Lindsay Olive Growers	34.7	0.05	100.0	0.0
36	Agripac	30.8	0.04	100.0	0.0
37	Zarda Brothers Dairy Inc.	29.0	0.04	100.0	100.0
38	Citrus Central Inc.	28.2	0.04	100.0	100.0
39	Cenex	0.2	0.00	100.0	0.0
	Total	73,649.6			

Note: % TV is the percent of the company's total media advertising spent on television. % SIC 20 is the percent of the company's total media advertising spent in food processing.

Source: Leading National Advertisers Inc., 1987.

TABLE 4.15 Media Advertising by Cooperatives in Food Processing, 1967

Rank	Company	Total ($000)	% of Cooperative's Total Advertising	% TV
1	Ocean Spray Cranberries Inc.	4,932	24.179	90.63
2	Allied Grape Growers	3,749	18.379	91.12
3	North Pacific Canners & Packers Inc.	3,587	17.585	97.41
4	National Grape Co-operative Association Inc.	2,749	13.477	100.00
5	California Canners & Growers	1,821	8.927	77.48
6	Sunsweet Growers Inc.	1,152	5.648	93.66
7	Land O'Lakes Creameries Inc.	1,010	4.951	4.36
8	California & Hawaiian Sugar Refining Corp.	634	3.108	85.33
9	Atlanta Dairies	619	3.035	59.94
10	Gold Kist Inc.	382	1.873	0.00
11	Arkansas Rice Growers Co-operative Assoc.	379	1.858	70.71
12	Diamond Walnut Growers Inc.	286	1.402	0.00
13	American Crystal Sugar Company	204	1.000	50.49
14	Sunkist Growers Inc.	191	0.936	84.82
15	Tree Top Inc.	187	0.917	21.39
16	Sun-Maid Raisin Growers of California	186	0.912	76.88
17	Dairymens Co-operative sales Assoc.	123	0.603	0.00
18	Guild Wine Company	93	0.456	0.00
19	Western Farmers Assoc.	77	0.377	0.00
20	California Almond Growers Exchange	59	0.289	0.00
21	Sioux Honey Association	58	0.284	25.86
22	Consolidated Olive Growers	47	0.230	0.00
23	Norbest Turkey Growers Association	43	0.211	0.00
24	Lehigh Valley Dairy Cooperative	37	0.181	0.00
25	Lakeville Dairies	27	0.132	100.00
26	Tillamook Cheese & Dairy Association	23	0.113	0.00
27	Challenge Cream & Butter Association	22	0.108	0.00
28	Consolidated Dairy Products Company	9	0.044	0.00
29	Missouri Farmers Association Inc	8	0.039	100.00
30	Roberts Dairy Company	5	0.025	100.00
31	Roquefort Association Inc	4	0.020	0.00
	Total	22,703		

Note: % TV is the percent of the company's total media advertising spent on television. 1967 advertising does not include non SICs 20 and 21 advertising, hence advertising for fresh farm products, e.g., oranges, are ignored.
Source: Leading National Advertisers Inc.

television advertising, but by 1987 only four of the 39 agricultural cooperatives failed to use some television advertising.

The above rankings have been done on the basis of total advertising but since cooperatives are not as large and diversified as the largest food and tobacco companies they might rank higher if the comparisons were done on the basis of advertising-to-sales ratios.   The calculation of such ratios is fraught with difficulties. First, our advertising totals are for U.S. media only and limited to food and tobacco products. Thus the companies' sales data must be similarly restricted to domestic sales of food and tobacco products. Such sales data are not easily available and confidentiality concerns impede firms from releasing such data.   However, a useful start is the annual list of the top 100 food processing companies done by *Food Processing* magazine in their December issue. They attempt to gather both total sales and food sales using the Census SIC system, but international food sales are included. Also, they do not include tobacco products. Nevertheless, advertising-to-sales (A/S) ratios were calculated using the sales data from the 1988 *Food Processing* list supplemented by estimates to add in domestic tobacco sales. No attempt was made to limit the food sales data to U.S. sales only, thus for firms with large international food sales the A/S ratio is substantially understated. For example, over half of the Coca-Cola company's sales are from overseas so its estimated A/S of 2.25 percent should be at least doubled to 4.5 percent if foreign sales were omitted.

Given these limitations, the advertising-to-sales ratios for the top 25 advertisers in 1987 were calculated and listed in Table 4.16. The A/S ratios vary from a low of 1.47 percent for Kraft Inc. to a high of 13.11 percent for Wrigley. Nine of the A/S ratios exceed 3 percent, the value picked by Brandow as marking excessive and wasteful advertising levels. In contrast, the A/S ratios for the top 25 cooperatives were dramatically smaller (Table 4.17). Hence, even after controlling for firm size, cooperatives do not advertise as intensively as the noncooperative firms. None of the cooperatives has an A/S exceeding 3 percent and only two have an A/S of 2 percent or more. Most are well below one percent. The highest, 2.6 percent, was for Ocean Spray which was also ranked number one on the basis of total advertising as well. In terms of food sales, Land O'Lakes was the largest of the leading agricultural cooperative advertisers, but even as the fifth largest cooperative advertiser its A/S ratio was only 0.33 percent.   Not only do cooperatives spend dramatically less on advertising than the leading noncooperative firms, they also have lower advertising-to-sales ratios.

Advertising by Cooperatives in Specific Food Industries

Although cooperatives are not among the leading advertisers in the food system, there are some processed food industries where cooperatives hold significant, even dominant, shares of a specific industry's total advertising. In some of the more narrowly defined product categories that still correspond to a well-defined economic market, like honey or raisins, cooperatives account for the majority of the advertising. However, the total advertising in these

TABLE 4.16 Advertising-to-Sales Ratios for the Leading Advertisers in Food and Tobacco Processing, 1987

Rank	Company	Comany Total ($000)	A/S Percent
1	Philip Morris Companies Inc.	769,772.9	3.50
2	RJR Nabisco Inc.	439,313.9	3.14
3	Anheuser-Busch Cos Inc.	325,465.8	4.07
4	Kellogg Company	309,008.9	8.15
5	General Mills Inc.	263,425.0	7.02
6	Mars Inc.	203,544.5	N/A
7	Pepsico Inc.	174,923.2	2.40
8	Coca-Cola Company	171,662.2	2.25
9	Nestle Sa	164,886.9	2.86
10	Kraft Inc.	145,139.1	1.47
11	Procter & Gamble Company	136,576.6	4.61
12	Quaker Oats Company	133,101.6	2.84
13	Campbell Soup Company	129,571.9	2.66
14	Unilever NV	124,408.1	N/A
15	Ralston Purina Company	124,147.5	2.71
16	Wrigley WM Jr. Company	102,379.1	13.11
17	American Dairy Association	93,987.1	N/A
18	BCI Holdings Corporation	86,342.0	2.15
19	Coors Adolph Company	84,656.8	7.70
20	Heinz HJ Company	84,441.6	1.70
21	Grand Metropolitan PLC	78,014.5	N/A
22	Hicks & Haas	76,115.6	N/A
23	Gallo E & J Winery	63,322.3	N/A
24	Hershey Food Corporation	62,406.8	3.35
25	Seagram Company Ltd.	61,939.7	2.82

Note: Company sales include foreign sales whereas only domestic advertising expenditures are included.

N/A = not available.

Source: Leading National Advertisers, Inc., 1987 and *Food Processing*, December 1988.

categories is often quite small and it is also in these markets where associations often fund nonbrand advertising as well.

We classified the 1987 advertising data for all food and tobacco products, including the unprocessed food industries, into meaningful economic product markets. Usually this amounted to using the five-digit product class but in some cases the more narrow seven-digit product was more appropriate and in others the four-digit industry was sufficient. The resulting set of economic industries were ranked by the percent of total advertising in the industry that was done by cooperatives, net of all nonbrand specific advertising done by associations and similar industry groups. In Table 4.18 all industries where the share of brand advertising held by cooperatives was one percent or more are listed in

TABLE 4.17   Advertising-to-Sales Ratios for the Leading Agricultural Cooperative
Advertisers in Food and Tobacco Processing, 1987

Rank	Company	Company Total ($000)	Percent of Total
1	Ocean Spray Cranberries Inc.	20,408.7	2.61
2	Sun-Diamond Growers of California	12,150.0	2.03
3	Agway Inc. (incl. HP Hood & C Burns)	7,882.0	0.96
4	California Almond (Blue Diamond)	5,110.6	1.12
5	Land O'Lakes Inc.	4,627.8	0.33
6	Sunkist Growers Inc.	3,890.3	0.46
7	Alexander & Baldwin Inc. (C & H Sugar)	3,677.6	0.63
8	Guild Winery & Distillers	3,502.0	N/A
9	National Grape Cooperative Association	2,641.6	0.86
10	Tri-Valley Growers	1,587.4	0.23
11	Tree Top Inc.	1,565.4	N/A
12	Farmland Inds Inc.	1,308.4	0.22
13	Citrus World Inc.	986.4	0.37
14	Gold Kist Inc.	911.2	0.11
15	Riceland Foods Inc.	353.0	0.06
16	Sioux Honey Association	336.7	N/A
17	United Dairymen/Arizona	332.8	N/A
18	Darigold Inc.	315.7	0.05
19	Coble Dairy	272.1	N/A
20	Challenge Dairy Products	263.2	N/A
21	Golden Guernsey Dairy Cooperative	183.5	N/A
22	Tillamook County Creamery Association	151.1	N/A
23	Cream O Weber Dairy Company	146.4	N/A
24	Prairie Farms Dairy Company	112.3	0.02
25	Dairymen Inc.	102.1	0.01

Note: Total advertising includes advertising done in SICs 01 and 02. Company sales
include foreign sales whereas only domestic advertising expenditures are included.
N/A = not available.
Source: Leading National Advertisers, Inc., 1987 and *Food Processing*, December 1988.

descending order of the cooperatives' share. Some duplication exists in the table
when the seven-digit product level is used as the appropriate economic market
because the five-digit product class is also given to show the implications of not
using the more narrow product level of detail. For example, both seven-digit
product markets for prunes and raisins are given separately but the five-digit to
which they belong is also given, SIC 20343, dried and dehydrated fruits and
vegetables. If one only used the five-digit product class level of detail, some of
the significance of agricultural cooperatives would be hidden.

In canned cranberries and sauce, SIC 2033128, the Ocean Spray cooperative
did all of the branded advertising and no association advertising was done.

TABLE 4.18 Share of an Industry's Brand Advertising Held by Agricultural Cooperatives, 1987

Cooperatives SIC	Name	Total Media Advertising ($000)	Total A/S[a] (percent)	Total Advertising by Cooperatives ($000)	Total Advertising by Associations ($000)	Share[a] (percent)
2033128	Canned cranberries and sauce	116.2	0.11	116.2	0.0	100.00
2034315	Prunes	6,136.7	1.25	3,325.7	2,811.0	100.00
2034313	Raisins	11,852.2	1.56	5,139.6	6,640.3	98.61
0173	Tree nuts	4,551.3	N/A	4,421.0	6.1	97.27
0174	Citrus fruits	5,382.1	N/A	3,934.2	1,225.9	94.66
2099G25	Honey, blended and churned	374.4	0.47	336.7	0.0	89.93
20210	Butter	18,088.5	0.14	2,051.9	15,785.3	89.09
20343	Dried and dehydrated fruits and vegetables	20,454.3	0.69	8,465.3	9,752.0	79.10
20623	Cane and beet refined sugar	6,461.7	0.11	3,677.6	1,706.9	77.34
0252	Chicken eggs	2,892.1	N/A	61.6	2,792.2	61.66
2026245	Sour cream	1,331.4	0.00	417.3	26.2	31.97
20262	Packaged fluid milk and related products	44,800.3	0.05	1,473.6	39,912.3	30.15
2048	Prepared feeds, n.e.c.	323.2	0.00	74.2	0.0	22.96
2033A,B	Canned fruit juices, nectars, and concentrates	105,516.3	2.96	19,342.4	8,036.6	19.84
2033136	Canned olives, incl. stuffed	906.5	0.18	101.0	397.3	19.84
20323	Canned dry beans	7,803.4	0.87	1,532.3	0.0	19.64
20680	Nuts & seeds (salted, roasted, cooked or blanched)	16,478.1	0.80	2,885.3	0.0	17.51

(continues)

TABLE 4.18 *(continued)*

Cooperatives SIC	Name	Total Media Advertising ($000)	Total A/S[a] (percent)	Total Advertising by Cooperatives ($000)	Total Advertising by Associations ($000)	Share[a] (percent)
20331	Canned fruits, except baby foods	24,016.8	1.13	2,942.4	397.3	12.46
20866	Noncarbonated soft drinks	39,985.2	1.65	4,587.6	0.0	11.47
20371	Frozen fruits, juices, ades, drinks & cocktails	32,826.1	0.96	2,404.0	5,621.2	8.84
20332	Canned vegetables except hominy and mushrooms	10,716.6	0.47	817.9	0.0	7.63
20263	Cottage cheese	3,471.8	0.43	178.5	544.8	6.10
20513	Sweet yeast goods, except frozen	3,475.8	0.35	206.6	0.0	5.94
20118	Canned meats	13,916.6	1.06	820.3	0.0	5.89
20338	Jams, jellies, and preserves	12,540.5	1.89	592.0	0.0	4.72
20151	Young chickens	20,994.9	0.28	911.2	160.9	4.37
20792	Margarine	49,941.3	4.28	2,105.9	0.0	4.22
2035435	Pourable dressing	19,991.0	4.05	676.8	0.0	3.39
20840	Wines, brandy, and brandy spirits	224,815.2	7.10	3,502.0	2,304.2	1.57
20354	Mayonnaise, salad dressing, and sandwich spreads	46,138.8	2.03	676.8	0.0	1.47
20240	Ice cream and ices	104,054.5	2.35	1,203.8	5,536.5	1.22
20223	Natural cheese, except cottage cheese	61,438.4	0.40	274.1	35,641.4	1.06

[a] Excludes association advertising from the industry total.
Source: Leading National Advertisers Inc., 1987.

The total amount of advertising was minor, only $116,200 and represented less than one percent of Ocean Spray's 1987 media advertising and created a mere 0.11 percent A/S ratio for the entire industry. Such a low A/S is characteristic of low levels of product differentiation in an industry.

Cooperatives also did all the brand advertising for prunes and the amount was more substantial, a little over $3.3 million of branded advertising, which amounted to an industry branded A/S ratio of 1.25 percent. Associations accounted for another $2.8 million. Cooperatives did almost all of the brand advertising for raisins with Sun-Diamond having spent $5.1 million or 98.6 percent of all branded advertising in support of the SunMaid brand. Associations spent more, $6.6 million, to support the raisin industry in general without mentioning specific brands.

Cooperatives were responsible for over 90 percent of all brand advertising for tree nuts and citrus fruits with both having about $4 million in brand advertising. Associations spent another $1.2 million in citrus fruits but almost nothing for tree nuts. Cooperatives accounted for nearly 90 percent of branded honey and butter advertising. Honey is a small industry and even though the total advertising was only $0.3 million it had a higher A/S ratio, 0.5 percent, than was found in the butter industry, 0.14 percent, even though cooperatives spent over $2 million dollars advertising branded butter. Associations spent nothing in the honey market but spent nearly $16 million in support of the butter industry.

Of the 10 product categories where cooperatives accounted for over 50 percent of the branded advertising, only two, butter and refined sugar, would be standard industries in the food processing sector (SIC 20). Three of the ten (tree nuts, citrus fruits, and chicken eggs) are classified outside of food processing. Four, including the top three product markets, required more detailed seven-digit data than are typically used in research because the Census only provides limited data at this level. Prunes and raisins are part of SIC 20343 (dried and dehydrated fruits and vegetables), which is also given in Table 4.18 as a standard five-digit product class. The effect of not separating out prunes and raisins from SIC 20343 reduced the cooperative's share of the branded advertising to 79 percent and lowered the industry's A/S ratio to only 0.7 percent, whereas it was over 1.2 percent for each when separated. The five-digit SIC 20343 is an overly broad classification as it includes noncompeting products. Although one could argue whether raisins and prunes belong in the same economic market, no one would suggest that they both belong with dehydrated potatoes which is also classified in SIC 20343.

Sour cream is a seven-digit product that is part of the five-digit SIC 20262, packaged fluid milk and related products, as was explained in Table 4.1. Both are in Table 4.18 and in each the percent of the brand advertising done by cooperatives was a little over 30 percent. However, associations spent nearly $40 million dollars in support of the broader five-digit group of products and rarely advertised for the more narrowly defined sour cream dairy products. Canned olives is another example of the usefulness of examining seven-digit products. Cooperatives spent about $100,000, or almost 20 percent of the branded advertising for canned olives, and associations spent nearly four times that amount. Canned olives is a subset of the five-digit SIC 20331, canned fruits. In the broader canned fruits industry cooperatives accounted for 12.5 percent of brand advertising and the only association advertising was done by the olive industry. The canned olives industry had an A/S ratio of only 0.2 percent, characteristic of low product differentiation, whereas the broader canned fruits industry had an A/S ratio of 1.1 percent.

Even though in a few product markets cooperatives dominated the brand advertising, the total amount of advertising done was minimal. The total amount of branded advertising accounted for by the industries listed in Table 4.18 represented only 13 percent of all branded food and tobacco advertising in 1987, but accounted for nearly 95 percent of all the advertising done by agricultural cooperatives. In only 19 product markets (and this double counts SICs 20331, 20343 and 20262 since they have seven-digit SICs also on the list) out of over 200 food and tobacco markets examined did the cooperatives' share of branded advertising exceed 10 percent of the total. Of these only one was an industry characterized as a moderate to large advertiser. That one industry was SIC 2033A,B (SIC 2033A and 2033B were combined), canned and fresh fruit juices, nectars and concentrates, where the industry spent $105.5 million on advertising, with $97.5 million on branded advertising for an A/S ratio of nearly 3 percent. Nevertheless, even this industry is not among the leading industries in advertising expenditures. For example, the breakfast cereals industry was the second largest advertiser behind the beer industry in 1987 and it spent over six times as much, $656.1 million, on branded advertising in 1987 and had an industry A/S of nearly 13 percent.

Three main conclusions emerge from Table 4.18. First, even though a few product markets had cooperatives dominating the brand advertising, the amounts were relatively small and the advertising intensity of the industry was small, usually less than 1 percent of sales, indicating low levels of product differentiation. Second, where the cooperatives did advertise branded products was also where associations spent the most to support the industry without regard to brands. The sum of the association advertising given in Table 4.18 represented 60 percent of all food related advertising by associations and that omits substantial amounts by associations that were not specific to a four- or five-digit SIC but were supportive of all the industries within a three-digit minor group (e.g., American Dairy Association's general dairy advertising). This implies that the farmers who may belong to a cooperative, but clearly belong to the industry, were spending greater amounts on nonbrand advertising than on brand advertising. Lastly, researchers interested in agricultural cooperatives and the food processing system need more detailed data than is available through standard Census data. The Census provides data on 53 food and tobacco industries (four-digit SIC level) and 160 product classes (five-digit SIC level). Very little aggregated data are published at the seven-digit product level and no product market concentration data or sales of individual firms can be obtained because of confidentiality protection. Researchers need greater access to private data series like that available from Information Resources, Inc. (IRI). These data are expensive but give price and sales data for individual brands gathered from retail scanner data. Such data could be matched with the LNA advertising data to provide a better understanding of the food processing sector in general, and in particular, how cooperatives fare within the sector.

In a separate study, Willis and Rogers prepared a data set on individual companies in 60 U.S. processed food and tobacco industries in 1987. The study used data from the Census of Manufactures and LNA, but also used the now discontinued brand level data from SAMI to calculate brand and firm market shares within each of the 60 industries (for details see Willis, 1992). These data allow a comparison between industries where agricultural cooperatives are leading competitors and those without a substantial cooperative presence. The measurement of a significant cooperative presence was done in two different ways. First, if any one of the four largest branded sellers in an industry was a cooperative, then that industry had a significant cooperative presence. The other method used a 1982 special tabulation of census data (see Rogers and Marion, 1990 for details) and classified industries into those where the 100 largest agricultural cooperatives collectively held at least a 3 percent share of the industry's value-of-shipments and those where it was less than 3 percent.

In twelve of the 60 industries at least one cooperative was among the top 4 sellers of branded products and the remaining 48 industries did not have a cooperative among the top 4. The two groups of industries differed in several important market characteristics (Table 4.19). The industries with cooperatives among the leaders tended to be slightly smaller, have lower market concentration, lower advertising intensities, higher levels of private label sales (those without a manufacturer brand identification, usually the retail store's brand), and lower price-cost margins as measured by either a Census approach (see Petraglia and Rogers, 1991) or by an average brand price minus the average private label price (see Willis, 1992). In addition, the industries with a significant cooperative presence tended to have their market leader face weaker rivals, as measured by market shares of firm one compared to firms ranked number two, three and four.

Essentially the same results emerge if the two groups are formed by whether the industry had at least three percent of its 1982 value-of-shipments held by cooperatives or not. This approach split the 60 industries more evenly, 25 had a significant cooperative presence and 35 did not. The comparisons of the mean characteristics repeat those already discussed based on the alternative measure of a significant cooperative presence. For this paper, the main point

TABLE 4.19 Comparison of Means of Industry Characteristics by Cooperative Involvement, 1987

Variable	All Industries	Cooperative in Top 4		Cooperatives' Market Share	
		Yes	No	≥ 3%	< 3%
N	60	12	48	25	35
Vos87	2,846.1	2,417.5	2,942.3	2,345.6	3,203.5
AS87	2.6	0.8	3.0	1.1	3.6
CR487	60.3	48.6	62.9	50.2	67.5
HHI87	1,597.1	1,095.6	1,709.6	1,067.9	1,975.0
MS1	34.3	31.3	35.0	30.6	37.0
MS2	16.4	8.4	18.2	12.0	19.5
MS3	10.1	5.9	11.1	7.6	11.9
MS4	5.4	4.4	5.7	4.5	6.1
PL	16.0	32.4	12.3	27.2	8.0
PCM1	34.5	24.7	36.9	29.0	38.4
PCM82	30.4	17.7	33.6	20.0	37.8
%COOP	7.0	24.2	3.1	15.9	0.5

where: N   = number of industries
Vos87 = 1987 value-of-shipments in millions of dollars
AS87 = 1987 advertising to sales ratio in percent
CR487 = 1987 four-firm concentration ratio in percent
HHI87 = 1987 Hirschman-Herfindahl index
MSi = brand market share of firm i
PL = percent of the industry's retail sales accounted for by private label
PCM1 = percentage average brand price exceeds average private label price
PCM82 = 1982 price-cost margin from Census (see Petraglia and Rogers, 1991 for details)
%COOP = 1982 percentage of the industry held by the 100 largest agricultural cooperatives (again see Petraglia and Rogers for details)

Source: Michael Willis and Richard Rogers, Department of Resource Economics, University of Massachusetts, 1992.

is the much higher advertising intensities for the group of industries without a significant cooperative presence. The A/S ratio for the group without a significant cooperative presence was over 3 times that of the ratio found in the group with cooperatives. However, this does not prove that cooperatives lower industry advertising levels. The results are also consistent with the fact that those industries where cooperatives are among the leaders have lower price-cost margins, lower concentration and are more commodity oriented.

To determine if cooperatives do spend less than their noncooperative rivals when all else is held constant is a difficult research task given data availability. As a preliminary test, we added the cooperative presence variable to a regression model that explains the industry's A/S as a function of price-cost margins, concentration, variance in leading firm's market shares, private label's share and some control variables (see Willis, 1992). The same 60 industries included in Table 4.19 formed the data set. The results showed a negative estimated coefficient on the cooperative presence variable but it was not statistically significant, hence we cannot support the hypothesis that industries with a major cooperative presence will have lower advertising intensities, all else equal.

We continue to develop additional data to test this hypothesis using brand level data from both the SAMI and IRI data, as well as from other trade sources. In broilers, for example, the market leader in 1989 was Tyson Foods with a 20 percent share of the national broiler market. Second was ConAgra with an 8.5 percent share. Gold Kist, an agricultural cooperative, was third with a 6.9 percent share and Perdue Farms was fourth with a 5.8 percent market share. In terms of the share of branded advertising for broilers though, Perdue was first with a 28.4 percent share of the advertising dollars spent in 1989, followed by Tyson with 24 percent and ConAgra with a 21.3 percent share. Gold Kist was a distance fifth with a 4.1 percent share of broiler advertising (see Rogers, 1993 in *Industry Studies*, edited by Duetsch). Unlike the red meat industry which spends heavily on industry wide advertising by associations, the broiler industry avoids industry-wide advertising in favor of individual companies advertising their brands.

A similar pattern exists for the U.S. turkey industry. In 1990 the four market leaders and their market shares were: Louis Rich (part of Philip Morris) at 8 percent, Swift (part of BCI) at 8 percent, Norbest, a cooperative, at 6.9 percent, and ConAgra at 6.4 percent. Their shares of branded turkey advertising were: 62 percent for Louis Rich, 20 percent for Swift, and only 0.3 percent for Norbest as well as ConAgra. In 1991, ConAgra bought Swift so it would be interesting to recheck the market shares and advertising shares after the merger, but Norbest is still likely to be an insignificant advertiser.

The olive industry is not one characterized by heavy brand advertising, the A/S ratio is only 0.1 percent for the industry. In 1987 associations spent nearly $400 thousand whereas branded olive advertising amounted to just half that amount. Campbell Soup had the largest branded olive market share with a 35 percent share and two cooperatives held positions two and three. Lindsay was the number two branded olive company with a 17.5 percent share and Tri-Valley was third with a 4.6 percent share. Although Campbell Soup spent the most money, $78.8 thousand, it had an A/S ratio of only 0.13 percent and Tri-Valley spent $66.3 thousand for an A/S ratio of 0.8 percent. The only other branded advertiser was Lindsay who spent $34.7 thousand for an A/S ratio of 0.11 percent. None of these three could be called substantial advertisers and the industry is not characterized by much product differentiation. In 1987 over 36 percent of the retail sales were in private label store brands or generic labels.

The SunMaid agricultural cooperative dominates the raisin industry. The SunMaid brand had about a 56 percent market share of the retail market in 1987. The brand was supported by over $5 million in advertising and that amounted to an A/S ratio of 3.7 percent based on retail sales, the largest A/S ratio we have observed yet for a branded food product marketed by an agricultural cooperative. Only two other raisin companies advertised their brand of raisins each spending about $35,000, but it was the smaller company, National Raisin Company, that had the larger A/S ratio of 1.24 percent for its Champion brand even though it had only a 1.2 percent brand market share. The raisin industry is clearly dominated by the leading firm, an agricultural cooperative, but the industry does not have strong product differentiation. Over 30 percent of 1987 retail sales were for private label store brands or generic labels. Also, the industry's association advertising exceeded that of the total branded advertising, spending $6.6 million in 1987.

Another agricultural cooperative, Sioux Honey Association, is number one in the honey market, with a retail market share of over 30 percent in 1987. The cooperative spent $337,000 supporting its Sue Bee brand, which gave a retail-based A/S ratio of 1.2 percent. Only one other company recorded any brand advertising and it was a small amount, $23,300, or an A/S ratio of 0.3 percent. The honey industry also has weak product differentiation and the industry had over 35 percent of retail sales made by private label store brands or generic labels but no association advertising was done in 1987.

At least three agricultural cooperatives operated in the retail sugar industry in 1987, but only one, California and Hawaiian Sugar (linked to Alexander Baldwin), advertised its brand of sugar. Although it was only the second largest sugar company based on its retail brand market share of nearly 10 percent, it outspent all others and accounted for over 80 percent of the total branded advertising in 1987. It spent $3.7 million for an A/S ratio of 3.1 percent. The market share leader was Amstar and it spent only $0.3 million for an A/S ratio of only 0.2 percent. The sugar industry has weak product differentiation also, with private label store brands and generic labels accounting for over 60 percent of the retail market share. Associations also spent nearly $2 million advertising in support of the industry in 1987.

Several agricultural cooperatives sell a branded butter product at the retail level. Land O'Lakes dominated the industry with a 33 percent market share of retail sales in 1987. It also dramatically dominated the branded advertising, accounting for over 84 percent of the total, but this amount, nearly $2 million, resulted in an A/S ratio of only 0.7 percent. The number two company was Kraft, now part of Philip Morris Companies, with a retail market share of 6.3 percent. It did not advertise its branded butter at all in 1987. Again the butter industry is not a differentiated market and private label store brands and generic labels accounted for nearly 45 percent of retail sales in 1987. Also, industry associations were the biggest advertisers spending nearly $16 million in 1987 in support of the butter industry.

In contrast to the butter industry where agricultural cooperatives dominate an industry characterized by little product differentiation and substantial association advertising, the margarine industry is dominated by noncooperatives, had no industry association advertising, but spent a substantial amount on brand advertising, nearly $50 million in 1987, and had only a modest share of retail sales made by private label store brands and generic labels (12.3 percent). The market leader was Unilever with a 31.3 percent retail market share and it spent over $26 million advertising its brands in 1987 for an A/S ratio of 5.6 percent. The second largest firm, RJR Nabisco with a 26.2 percent market share, spent over $16 million on its branded margarine for an A/S of 4.2 percent. Kraft was third with a 20.2 percent market share and spent $4 million on its brands for an A/S ratio of 1.3 percent. Land O'Lakes was the only cooperative among the top 4 firms with a market share of 3.1 percent. It spent $600,000 on its branded margarine for an A/S ratio of 1.3 percent.

The differences between the butter and margarine industries offer useful insights into advertising strategies. Whenever the industry has product differentiation potential, investor-owned firms spend large sums promoting their individual brands rather than contributing to industry-wide association advertising. The resulting advertising levels are much higher than found where the reverse is true. Even though association advertising levels have grown substantially in recent years—to the point where some within the industries are calling on their industries to stop the practice—the amount of advertising still does not rival that spent by the leading food and tobacco processing firms in industries where product differentiation can be built and maintained through massive advertising expenditures.

The evidence gathered to date does show that in a few markets cooperatives have similar advertising strategies to their noncooperative rivals and may even earn a price premium in these markets. However, these are the exceptions. On average, where an agricultural cooperative has a dominant position it tends to be in a market without heavy advertising rivalry, is less concentrated and sales of unbranded products comprise a large part of the market. In addition, when a cooperative does hold a dominant position in a market it is often left unchallenged by the large marketing firms that dominate much of the food system.

Summary

This review of advertising by agricultural cooperatives has found that their collective share of brand-oriented food advertising was unchanged over a 20-year period and remained a smaller percentage than their share of processed food sales. The agricultural cooperatives held their greatest advertising shares in food processing markets that had low value-added to sales ratios, low product differentiation, were commodity based, and had a high proportion of unbranded sales, even in retail stores. In addition, those markets were not highly concentrated and were not dominated by the twenty largest food processors who prefer differentiated products and advertising rivalry to direct price competition.

There are several reasons for this. An agricultural cooperative is usually an extension of the farm enterprise, governed by homogeneous boards with most if not all of the board members having farm production backgrounds, and the primary objective of cooperatives has been to assure their farmer/members a market their output. This has often led to a production orientation rather than a marketing outlook. The additional volume that cannot be sold under the cooperative's retail brand name is likely to be sold as private label store brands or generic labels rather than kept off the market. Agricultural cooperatives usually operate in first-stage processing that is characterized by low value-added products and are often undercapitalized limiting further extensions. In addition, there are substantial barriers to mobility from expanding the unbranded slice of a market to the national brand strategic group that is often dominated by the huge marketing-oriented firms that comprise the twenty largest food and tobacco processors.

Although a clear picture has emerged showing that agricultural cooperatives advertise less than noncooperatives, both in total amounts and intensity levels, we cannot conclude that the lower emphasis on advertising is the result of being an agricultural cooperative or just related to the market characteristics that cooperatives operate in. Of course, the two are likely to be interrelated making research difficult to discover the true causal linkages. Numerous examples exist where noncooperatives in commodity-oriented markets with little to no product differentiation launched substantial advertising campaigns and achieved some success. Perdue's aggressive advertising of its broilers along with industry leader Tyson's continual attempts to add value to the basic product are prime examples supporting that position. Few examples exist where agricultural cooperatives have tried such a marketing approach. Whether these are just antidotal examples or are part of a general explanation await further research.

References

Boynton, Robert D. 1982. *A Comparison of the Advertising Expenditures of Cooperative and Non-cooperative U.S. Food Processors.* Indiana Agricultural Experiment Station, Bulletin No. 379.

Brandow, George E. 1977. Appraising the Economic Performance of the Food Industry, in *Lectures in Agricultural Economics.* Washington, D.C.: USDA.

Connor, John M., Richard T. Rogers, Bruce W. Marion, and Willard F. Mueller. 1985. *The Food Manufacturing Industries — Structure, Strategies, Performance, and Policies.* Lexington, MA: Lexington Books.

Dorfman, Robert and Peter O. Steiner. 1954. Optimal Advertising and Optimal Quality, *American Economic Review.* XLIV(December): 826-836.

Kinnucan, Henry W., Stanley R. Thompson, and Hui-Shung Chang. 1992. *Commodity Advertising and Promotion.* Ames, IA: Iowa State University Press.

Petraglia, Lisa M. and Richard T. Rogers. 1991. The Impact of Agricultural Marketing Cooperatives on Market Performance in U.S. Food Manufacturing Industries for 1982. Food Marketing Policy Center, Research Report No. 12, University of Connecticut, Storrs, Connecticut.

Rogers, Richard T. 1982. Advertising and Concentration Change in U.S. Food and Tobacco Products, 1954 to 1972. Ph.D. Thesis, University of Wisconsin.

___. Broilers: Differentiating a Commodity. In *Industry Studies,* ed. L. Duetsch, 3-32, Englewood Cliffs: Prentice Hall.

Rogers, Richard T. and Bruce W. Marion. 1990. Food Manufacturing Activities of the Largest Agricultural Cooperatives: Market Power and Strategic Behavior Implications, *Journal of Agricultural Cooperation.* 5: 59-73.

Rogers, Richard and Randall Torgerson. 1988. Ag Marketing Co-op Participation Limited in Food Processing Industry, Farmer Cooperatives. 55(7).

Willis, Michael S. 1992. Leading Firm Heterogeneity as a Determinant of Advertising Intensity in Food and Tobacco Manufacturing, M.S. Thesis, University of Massachusetts.

5

Market Strategies in Branded Dairy Product Markets

Ronald W. Cotterill and Lawrence E. Haller

Dairy cooperatives face several strategic options, one of which is integrating forward into processing to market branded dairy products. This paper documents the extent of cooperative penetration into branded product markets and presents some rudimentary case study evidence on competitive strategies in those markets. We employ data for 51 local markets as well as national data from the Information Resources Inc. (IRI) "Supermarket Review" data base for 1988 and 1989. To our knowledge this paper is the first systematic examination of the position of cooperatives and investor owned branded dairy product marketers in local markets. We identify regional cooperative brands that do not rank high when looking at total national sales but have leading market positions in their chosen markets.

The next section of this paper provides information for the top 20 firms ranked by national sales and for all cooperatives that market one or more branded products in the following product categories: skim/low fat milk, whole milk, cottage cheese, butter, margarine, and ice cream. Although margarine is not a dairy product, we include it because it may be a close substitute for butter. Throughout this paper we assume that these product categories are relevant product market definitions for strategy analysis. For the present paper this is a workable assumption. However, a more detailed analysis may indicate that products such as skim/low fat milk and whole milk are in the same product market.

For each of the top 20 firms and other cooperatives we identify their 1989 national product category share, their average price for 1989, the number of local markets that the company is in, and the number of local markets where it ranks first, second, third, or fourth.

The third section of this chapter uses local market data on a few selected brands to analyze in graphic form how prices, quantities and category shares change over time in particular markets, and how a brand's price and category share vary for a particular time period across several local markets. These simple graphs provide considerable insight into brand marketing strategies. The last section contains conclusions.

National Market Position, National Price, and Local Market Positions

The IRI Supermarket Review data base uses scanner data collected from over 2400 supermarkets nationwide to estimate several economic variables for brands such as Land O'Lakes butter on a quarterly basis for 51 local markets. The graphic definitions of the local areas are illustrated in Figure 5.1. These local market areas range in population size from Boise, Idaho to the metropolitan New York area. Information Resources, Inc. and A.C. Nielsen are the only two companies that provide this type of data. Food manufacturers regularly use a more detailed version of these data (weekly reports for particular container sizes) for the daily operation of their brand marketing programs. Aggregation to quarters and brands may prevent analysis of very short run competitive dynamics. However, it should provide sufficient detail to track longer run strategic interaction.

Table 5.1 identifies total market private label sales and brand sales for the top 20 firms and all cooperatives in the skim/low fat product category. As is commonly known, private label volume dominates branded product movement. Private label volume accounts for 63 percent of skim/low fat national volume in 1989. The 1989 average price was $2.09 per gallon. Since not all skim/low fat milk was sold in gallon containers, and the price "per gallon" of milk sold in smaller unit sizes is generally higher, we have included the variable "units per gallon" to facilitate comparison of prices across companies and brands. Private label milk sales averaged 1.33 units per gallon. If all sales had been in gallons this variable would be 1 unit per gallon; if all sales were in half gallons it would be 2.0 units per gallon. Brands with higher units per gallon in Table 5.1, as expected, have higher average prices per gallon. A comparison of brand prices needs to control for differing units per gallon, if there is significant variation among brands.

Borden is the leading marketer of branded skim/low fat milk. Its share of national sales is only 2.9 percent and that share is distributed across seven brands, with the "Borden" brand capturing most sales (1.33 percent of national sales). The Borden brand has a $2.47 average price per gallon. This is well above the private label price. However, part of the brand differential is explained by the somewhat smaller unit size (1.51 versus 1.33 units per gallon). Borden Inc. sells its branded product in 27 of the 57 local markets and, based on sales of all its brands, it is the leading firm in 6, the second firm in 8, the third firm in 5, and the fourth firm in 4 local markets. Only in four out of its 28 markets does Borden Inc. rank below fourth.

Local category share ranking on a brand basis differs from company rankings because companies with large shares spread over several brands may have relatively low shares for individual brands. This is the case for Borden Inc. On a brand basis it occupies the leading market position in 5 rather than 6 markets. It sells its leading brand, Borden, in 18 local markets and it is the top brand in 2 markets. "Meadow Gold" is the leading brand in 2 markets and the "Lite Line" brand leads in another market.

Table 5.1 also indicates that after Borden Inc., Dean Foods and Philip Morris are distributed most broadly across markets. Each operates in 10 or more local markets. The largest cooperatively controlled firm in the skim/low fat category is Agway/Hood. It ranks number 8 nationally with sales in four local markets. Agway/Hood markets "Hood" and "Hood Nuform" brands and has relatively strong market position, being the number one brand market in one market and second in its three other markets.

Darigold Inc. ranks number 9 in national sales, and is the second largest cooperative. It sells in 2 IRI local markets and it is the leading seller of branded milk in each of them. The other two cooperatives in the top 20 are Highland Dairy Inc. (No. 12) and Prairie Farms Inc. (No. 19). Highland sells in 3 local markets, is number 4 in two and ranks fifth or higher in the other. Prairie Farms sells in 4 local markets and ranks second in one, fourth in another and fifth or lower in the other two.

Moving beyond the top 20 we have identified in Table 5.1, 14 other cooperatives operate in the skim/low fat product category. In the aggregate they operate in 23 local markets, rank first in 5, rank second in 6, rank third in 1, rank fourth in 7, and rank fifth or lower in only 2 markets. Thus, when one examines market share positions in local markets, cooperatives are considerably stronger than is indicated by their national product category shares.

Moving to the whole milk category, as reported in Table 5.2, one finds quite similar results. Borden Inc., Philip Morris and Dean Foods are again the large multi-market players. Nearly all of their local market operations rank in the top four. Three cooperatives are among the top 20 firms. These are Hiland Dairy Inc. (12), Flav-O-Rich Inc. (14) and Agway/Hood (18). These three cooperatives operate in 13 local markets and rank first in 1, second in 5, third in 3, fourth in 2, and fifth or lower in 2 local markets. Fourteen other cooperatives sell whole milk in 23 local markets and rank first in 7, second in 4, third in 2, fourth in 5, and fifth or lower in 5 local markets.

Table 5.3 examines the cottage cheese category. Private label volume again accounts for a significant share of category volume (40.5 percent). Average price for private label in 1989 is $1.01 per pound, and firms, on average, sell 1.27 units to distribute a pound of cottage cheese to consumers. As expected, the same large fluid processors appear in the top 20. Philip Morris/Kraft is the leading national firm with 6 brands accounting for 21 percent of national volume. Sealtest "Light n'Lively" is the leading brand with a 7.96 percent national share of cottage cheese volume. Its average price per pound is $1.36. This is $.35 higher than the private label price, and since Lite n'Lively is on average sold in larger containers than private label, this brand price differential may be somewhat understated. Philip Morris/Kraft sells cottage cheese in 40 of the 51 local market areas, and ranks first in volume in 22 markets, second in 7, third in 9, fourth in 1, and below fourth in only one local market.

Agway/Hood ranks third in national volume and is the largest cooperative processor. Its "Hood" brand, however, is sold in only six local markets, but it is the top ranking firm in five of them and number 3 in the sixth.

The other cooperatives in the top 20 are Darigold (9), Prairie Farms (12), Intermountain Milk Producers (14), Golden Guernsey Dairy (17), and Cabot Farmers' Cooperative Creamery (18). Together these firms operate in 15 local markets, rank first in 4, second in 1, third in 4, fourth in 5, and below fourth in only one local market. Cooperative cottage cheese processors that rank among the top 20 processors generally have achieved their size by building strong market positions in relatively few local markets. There are nine other cooperatives that sell branded cottage cheese. They operate in 16 local markets and rank first in 4, second in 2, third in 3, fourth in 3, and lower than fourth in 4 markets. Thus, all cooperatives report 37 brand positions in the 51 local markets and have a rank of fourth or higher in 32 cases. These cooperatives are leading brands in most of their local market areas.

Table 5.4 reports on the butter category. Again private label sales are a major competitive factor with 44 percent of the national market. The 1989 average price per pound for private label is $1.82 and on average 1.05 units are sold to distribute a pound of butter. In the branded product segment Land O'Lakes is the dominant market player. The cooperative has 31.4 percent of the national market, more than seven times the share of the second largest firm, Philip Morris (4.4 percent). The average price per pound for Land O'Lakes butter is $2.11 and is thus $.29 above the private label price. Land O'Lakes operates in all 51 local markets and ranks first in 38 markets, second in 8, third in 3, fourth in 4, and lower than fourth in only 1 market.

Unlike multi-market firms in milk and cottage cheese, Land O'Lakes has only two brands, with the "Land O'Lakes" brand accounting for more than 99 percent of its sales. Philip Morris/Kraft, the second largest seller of butter, offers three brands; however, its Breakstone brand accounts for over 90 percent of its butter sales. Borden, the third largest seller of butter, has five brands with "Kellers" and "Hotel Bar" accounting for 47 and 41 percent of total company butter sales respectively. Multiple brand strategies may not be as common in butter possibly because, to date, butter is a relatively homogenous product that has seen declining per capita consumption for health reasons. Perhaps the advances in fat substitutes and cholesterol removal technologies will offer options for new brands in the future.

Cooperatives are more common among the top 20 firms in butter than other dairy products industries. Besides Land O'Lakes, eight others rank in the top 20. Each of these firms' volume, however, is less than one tenth of Land O'Lakes' volume. In combination they sell branded butter in 33 markets and rank first in 7, second in 9, third in 6, fourth in 6 and below fourth in 5 local markets. Ten other cooperatives also market brands of butter. In combination they distribute in 13 local markets and rank first in 2, second in 3, third in 4, fourth in 2, and below fourth in 2 local markets. Butter sales in 1989 through supermarkets totaled approximately 310 million pounds. Margarine sales were much higher totaling 1,671 million pounds.

Table 5.5 reports product category position and price information for all 22 firms in the margarine and spreads category. Private label plays a much lower role here than it does in the dairy categories. Only 16 percent of margarine volume is private label. The 1989 average price per pound is $.49 and on average 0.84 units are sold to distribute a pound of margarine.

The brand structure of the margarine category is quite different than butter. The two leading firms, Unilever and RJR Nabisco, distribute several brands. As a company, Unilever ranks first in 25 local markets but ranks first in only 8 markets at the brand level. In the other 17 markets its leading position as a company comes from sales of two or more lower ranked brands. RJR Nabisco has two flagship brands, "Blue Bonnet" and "Fleischmanns", that each have 11.4 percent of the national market. Fleischmanns has a somewhat stronger position in local markets.

Philip Morris is the third largest firm with over 75 percent of its sales accounted for by its "Parkay" brand. Parkay is the number one brand in 20 local markets, more than any other brand.

Land O'Lakes is the largest cooperative in the margarine and spreads category. It ranks fifth in category sales, selling two brands. "Land O'Lakes" margarine is sold in 39 local markets but its brand share is fifth or lower in 33 of them. Its "Country Morning Blend" margarine sells in 41 markets and is never one of the top four brands. Thus, in the margarine product category Land O'Lakes is not a leading competitor.

Table 5.6 reports on the ice cream category. Private label accounts for 40.8 percent of national volume and the price per half gallon averaged $1.88 in 1989. Philip Morris/Kraft is the largest player and distributes its brands in 46 of the 51 local markets. On a company basis it ranks first in 18, second in 9, third in 6, fourth in 5, and below fourth in 8 local markets. The brand structure in ice cream is highly differentiated with four types: regular, all natural premium, all natural super premium and light (lower calorie) ice cream. The leading national brand is Breyers, an all natural premium ice cream, sold by Philip Morris. The price for Breyers, at $3.46 per half gallon, is more than a dollar above the price for the leading regular ice cream, Sealtest ($2.37) which is also sold by Philip Morris. Frusen Gladje is the Philip Morris super premium brand. Note that it sold only in pints (4 units per half gallon) and its 1989 average price was $8.30

per half gallon or $2.07 per pint. Two ice cream operations that market only super premium brands are in the top 20 firms. Haagen Dazs brand (No. 7) sold by Grand Metropolitan PLC is also sold only in pints and averaged $2.26 per pint. Ben and Jerry's (No. 16) is sold in pints at an average price of $2.21 per pint. Haagen Dazs is the most widely distributed super premium selling in 43 markets and ranking third in 2 markets and fourth in 5 markets. Ben and Jerry's is distributed in 20 markets and ranks below fourth in all markets.

Agway/Hood is the largest cooperative ice cream processor (no. 6). Its leading brand is "Hood," a regular ice cream, and it also sells two light ice creams, "Hood Light" and "NuForm." Agway/Hood distributes in seven markets and ranks first in 2, second in 2, third in 1, and below fourth in 2. Note that "Hood Light" sells at a premium to regular "Hood" ($2.24 versus $2.04 per half gallon).

Three other cooperatives rank in the top 20: Flav-O-Rich (no. 14), Prairie Farms (no. 17), and Darigold (no. 20). Moving beyond the top 20, ten cooperatives distribute brands of ice cream. When all cooperative operations are totaled they sell brands in 36 markets, rank first in 9, second in 9, third in 4, fourth in 3 and below fourth in 11 local markets. No cooperative sells super premium ice cream. In the premium category Flav-O-Rich sells "Rich and Creamy" in seven markets and Prairie Farms sells "Old Recipe" in 3 markets. Neither brand ever ranks in the top four.

Clearly cooperatives have not moved into the new product niches as rapidly as IOF ice cream manufacturers. Ben and Jerry's, a start-up firm in Vermont, has had spectacular growth with its innovative super premium ice cream. Haagen Dazs was a similar start up venture that was subsequently acquired by Pillsbury, which was acquired by Grand Metropolitan. This suggests that, at least in some cases, successful new brand development requires creativity and not large firm size.

Table 5.7 summarizes the market position of dairy marketing cooperatives in these five product categories. When one examines the column titled "national rank of cooperatives in top 20," the number of cooperatives ranges from three in the whole milk category to nine in butter. The total number of cooperatives in each category is relatively uniform ranging from 14 to 19 cooperatives. Thus, category-based joint marketing ventures or other forms of cooperation among cooperatives would require cooperation from relatively few cooperative organizations. The fact that cooperatives, and primarily Land O'Lakes, dominate the butter trade is well documented. There are 97 instances of cooperative distribution in the 51 markets, and a cooperative ranks first in 47 of the markets. Cottage cheese is a distant second in terms of cooperative penetration. Cooperatives have 37 instances of cooperative distribution of branded products in the 51 markets, and 13 first place positions.

If one totals market positions across the five dairy product categories and examines the percent penetration by cooperatives for each of the top four market positions, one obtains a crude measure of the relative position of cooperatives as a group. Cooperatives account for one third of the first place positions in these five categories, 20.8 percent of the second positions, 12.9 percent of the third positions and 17.6 percent of the fourth positions.

When combined with their extensive private label operations, which are undocumented in this paper, cooperatives are a significant competitive factor. Frankly we were somewhat surprised to learn that cooperative penetration into branded dairy product sales, beyond butter, is this extensive at the local market level. Low shares of national product movement do not translate into low market shares in local market areas. We turn now to an examination of particular investor owned firm (IOF) and cooperative brands to explore the significance of this new local market information for the formulation of marketing strategies.

Preliminary Case Studies of Related Brands

Except for related work by Haller (1993), previous quantitative analysis of branded food product manufacturers has exclusively employed aggregate national data on average profits, prices, volume movements, product category or more aggregate census category shares, and other structural variables to analyze strategy and performance issues. Examples include market structure-profit studies as reported in Cotterill and Iton (1993), brand-private level price difference studies such as Parker and Connor (1979), Wills (1985), and Connor and Peterson (1992) and conjectural variation studies such as Wann and Sexton (1993). As Rogers (Chapter 4) explained in the prior paper in this workshop, national advertising is important for many food products. It plays a central role in creating strong brand preferences and significant brand price premiums. The reported national average price differentials in Tables 5.1 through 5.6 of this paper affirm that leading brands in all product categories including the fluid milk categories, which is often regarded as an undifferentiated product, command premiums over private label prices.

The question that we can now analyze is how do leading brands compete in local market areas? The IRI Supermarket Review provides panel data for each brand; e.g., there are observations for the eight quarters of 1988-1989 across as many as 51 local markets. Rather than formally specify models and test them we will look at scatter plots of the data to see what they can tell us about competition and possible ways to model competition. This is primarily an inductive, case study approach. To do this effectively, however, it may be useful to briefly introduce some theoretical concepts to explain how different tests for market power are related and the conduct that each predicts we would observe in markets so that we have a framework for the discussion of diverse observed phenomena.

The residual demand approach estimates a demand curve for individual brands in differentiated product markets (Baker and Breshnahan, 1988). If the curve has negative slope the firm has power over price. The more inelastic the brand demand relationship, the more power the firm has. Here, we expand this concept in a fashion that helps to relate it to market share tests for power by introducing the concept of a followship demand curve. As illustrated in Figure 5.2, if all firms in an oligopolistic market raise and lower price together; i.e., follow each other's price, then each firm faces a followship demand curve and has a constant market share as prices fluctuate. The followship demand curve has negative slope because as all prices decline market demand for the product increases. This type of pricing conduct suggests tacit collusion or price leadership rather than price chiseling, rivalry, or other procompetitive pricing strategies. A firm's conduct is rivalrous if, as illustrated in Figure 5.2, a price cut from p_0 to p_1 results in an expansion in quantity beyond that necessary to sustain a constant share. The quantity increase $Q_0 Q_F$ is consistent with a constant share, and the second component of the observed quantity increase, $Q_F Q_A$, is indicative of rivalry. For the illustrated case the firm's increase in volume comes from capturing market share from other firms as well as increases in total market quantity demanded.

If rivalry is complete (perfect competition) the actual demand curve will be flat. Thus, the ratio of the slope of the actual demand curve to the followship demand curve ranges from zero (perfect competition) to 1 (perfect price coordination). This index of market power controls for the fact that different brands (firms) will have different residual elasticities due to different market shares and, thus, enables systematic comparison of the power index values across brands (firms).

In summary, if one observes a negative relationship between price and quantity (volume), the firm does have power over price. However, if the firm's observed negative price-volume relationship is consistent with a constant market share, i.e., there is no relationship between price and market share, then the conduct is collusive. Alternatively, if market share and brand price are negatively related then some interbrand rivalry is present.

This generalization of the residual demand analysis of pricing conduct predicts that market share and price are negatively related in rivalrous or competitive markets and not related in noncompetitive markets. How can one reconcile this with the general literature on oligopoly theory that predicts profit maximizing firms with larger market shares; i.e., more concentrated markets, may have higher prices? Following work by Harris (1988), we have been able to reconcile the two as follows. In a differentiated product oligopoly model a firm (brand) residual demand elasticity is a function of several variables including market share. As market share increases, the firm (brand) residual demand becomes more inelastic and the profit maximizing price increases. Intuitively, as a firm expands its share, it moves from a price taker, with infinitely elastic demand, towards a monopolist who faces the market demand curve. Thus, as illustrated in Figure 5.3, we would expect to find more inelastic residual demand curves, price followship, and higher prices in markets where market share is high. In low share markets the residual demand curve is flat (elastic), nonfollowship (rivalry) dominates, and prices are low. We would note that this explanation assumes constant production and distribution costs, that differentiation is costless, and that different levels of differentiation generate the observed share distribution. The underlying model does allow for relaxation of these assumptions and, if larger share firms enjoy economies of scale, the cost efficiency effect may affect or even dominate the power effect of large share, producing no share price relationship or a negative relationship. We refer readers to Cotterill (1993) for a more rigorous presentation of this model.

This theory suggests that examination of a brand's price-volume conduct across time (eight quarters) in a particular market will generally produce a negative relationship between brand price and volume because quarterly time series analysis should capture short run shifts in supply conditions that trace out a relatively stable demand curve. In cross section analysis one may find a positive relationship between share and price if the power effect dominates the cost effect, no relationship if they cancel each other, and a negative relationship if cost effects dominate.

We will analyze the following brands: Land O'Lakes and Crystal Farms butter; Imperial, Parkay and Land O'Lakes margarine; Breyers and Hood ice cream; and Borden and Deans skim/low fat milk. The price-volume, price-share, and occasionally the price trend over time will be analyzed for brands over eight quarters in particular local markets. Then we will examine the cross section scatter plots between a brand's 1989 average price and its 1989 product category share. One *caveat* is in order. All reported brand prices are shelf prices and are not adjusted for manufacturer coupon redemption. Some of the very high brand prices in particular markets or particular quarters may be due to coupon merchandising strategies.

The first product that we examine is butter in the Chicago retail market area. Figure 5.4 illustrates the relationship between quarterly price and volume for Land O'Lakes, Crystal Farms and private label butter. Land O'Lakes is the leading brand in Chicago and sells at a hefty premium over private label. Crystal Farms is the second brand in Chicago and sells at approximately the private label price. The scatter plots for Land O'Lakes and private label seem to identify demand curves with substantial slope. Assuming that they do identify demand curves, the question is do increased sales come from other firms (increased market shares) as well as from moving down the market demand curve or do they come only from the latter (constant market shares)? Figure 5.5 provides a provisional answer. It displays the scatter plots for price and shares. For Land O'Lakes, share is constant or possibly positively related to price.

Similarly the private label price changes do not produce large fluctuations in private label share. Price followship seems to hold in this market where the combined share of Land O'Lakes and private label butter exceeds 90 percent. Crystal Farms butter is a very marginal player in this market. Its prices are not related to its volume or share. The lack of a price volume relationship for Crystal Farms suggests that it behaves as a competitive fringe firm, regarding market price conditions as a given that it cannot influence in its marketing activities.

Moving to the margarine market in the Chicago retail area produces a different story. Since private label is not a major player in margarine markets we will ignore it. Figure 5.6 is somewhat messy but it illustrates the price volume relationships in Chicago for Imperial, Land O'Lakes, and Parkay margarine. Land O'Lakes is generally the highest priced, then Parkay and then Imperial. Their volumes span the similar ranges. Each brand seems to identify a negatively sloped brand demand curve. The three brand demand curves seem to identify a single demand curve. However, recall that these brands are being sold at the same time for three different prices in the local market, so this is not the case. Figure 5.7 looks at price-share scatter plots for these three brands. First, note that in combination these three brands account for only 50-60 percent of margarine sales. All three brands exhibit significant negative relationships between price and category share. In fact, when comparing figures 5.6 and 5.7, it appears that most of the added volume due to lower prices comes from share gains and not increased total sales of margarine in the market. Thus, the Chicago margarine market seems very rivalrous. This is a very different conclusion than we reached for Chicago butter.

Shifting now to cross section analysis of butter and margarine brand prices, Figure 5.8 displays the scatter plot for the 1989 average price for Land O'Lakes butter and its category share in the 51 local markets. The brand price does not appear to be significantly correlated with its share and the correlation is, if anything, slightly negative. Figure 5.9 is a similar scatter plot for Land O'Lakes margarine. Price for this brand does appear to be somewhat positively related to share. Thus, it appears that Land O'Lakes is following different geographic pricing strategies for these two products. This may be because of different cost-share relationships for the two products or because Land O'Lakes as a cooperative is pursuing a volume maximizing strategy for butter to move product and is pursuing a profit maximizing strategy for margarine to generate earnings for its dairy farmer members.

Figure 5.10 illustrates the price-share scatter plot for Philip Morris' "Parkay" margarine. Clearly, there is a negative relationship between brand price and local market share. Figure 5.11 is for Unilever's Imperial margarine. The possibility of a positive price share relation surfaces again for this brand. In summary, these cross sectional scatter plots suggest that pricing strategies at the brand level in differentiated markets can vary significantly among firms and possibly among products in a single firm.

Figure 5.12 illustrates the quarterly price-volume relationships for Breyers, Hood, and private label ice cream in the Boston retail market area. Breyers, the leading national brand, sells at a premium that primarily reflects the fact that it is an all natural premium ice cream. The Hood and private label products are regular ice creams. Note that Hood consistently sells at a premium to private label. The price-volume points for Breyers identify what seems to be a demand curve that is considerably less elastic than the Hood or private label demand curves. Thus, Breyers seems to have and exercise considerably more pricing discretion than Hood. Figure 5.13 displays the corresponding price share relationships. Breyers' market share is considerably less sensitive to price changes than Hood's or the private label products. Again, just as for margarine, this suggests that the ice cream market is segmented and that premium ice cream does not compete as directly with regular ice cream as it does with other premium ice creams.

Figure 5.14 examines the cross section relationship between Breyers ice cream price and share for the local markets that it operated in during 1989. Clearly, there is a negative share-price relationship. Share-related cost effects seem to dominate market power as a source of profits. For whatever reason, consumers benefit when Breyers has a large market share.

Our last case study is skim/low fat milk for Borden and Deans. Figure 5.15 is the quarterly price trend for two brands, "Borden" and "Deans", and private label for 1988-1989 in the Chicago retail market area. Note that Deans, the market leader, sells at a significant premium over "Borden" and private label, which are nearly identical. Figure 5.16 examines the relationship between price and volume, and price and share for the Borden brand. The price-volume scatter plot looks more like a supply than a demand curve. The price-share scatter plot also appears to have a positive slope. We have checked the data carefully to make sure there are no computational errors. At this point we have no explanation for this conduct which is very divergent from all other brands analyzed in this paper. The upward trend in prices over 1988-1989 seems to be due to the strong outward shift in demand for milk and stable supply conditions. Possibly this shift is due to nonprice merchandising. We would welcome alternative explanations that future research may provide.

Figure 5.17 is for Deans milk in Chicago and exhibits the same strong positive relationship between price and volume as Borden. However, the price-share relationship for Deans collapses into a vertical spike that suggests changes in Deans volume came from changes in total market volume, not from gains in share from other competitors. Note that Deans accounted for approximately 17 percent of product category sales, and Borden accounted for only 3.5 - 4.5 percent. Deans is the market leader and appears to have raised price in a strong enough fashion to limit its share gains. Borden raised price but not so strongly and, consequently, its market share expanded. Thus, we seem to have an example of a leader that wants to practice followship supply pricing (price leadership or collusion) and a much smaller rival that is content to chisel a little bit on the leader's intentions.

Private label skim/low fat milk behaves in a completely different fashion that is consistent with the conduct reported for all brands except Borden and Deans. Figure 5.18 reports price and volume and price and share on the same graph. Each suggests a strong negative relationship and, thus, a nonfollowship demand curve. Note that private label accounts for 51-58 percent of the market. When prices were higher in 1989, private label lost share, Borden gained share and Deans remained roughly constant. This seems consistent with the idea that the market leader tried to lead price up, the fringe brand chiseled by not raising price enough and gained share from private label brands that followed the leader.

Shifting to cross section share-price relationships, Figure 5.19 indicates that for the Borden brand there is little relationship between 1989 average price and 1989 average share across the 14 local markets that it supplies. Chicago is one of its lowest priced markets. Depending on how some of the extreme observations are explained by other variables in a more complex model, a significant positive or negative relation could, however, easily materialize. Figure 5.20 reports on the share-price relationships for the Deans brand in 10 local markets. There does appear to be a positive share-price relationship with Chicago being the largest share market and the third highest priced market.

Summary

Local market information from scanners as provided by Information Resources Inc. and A.C. Nielsen clearly provide researchers with the opportunity to establish powerful new insights on competitive strategy, market power and efficiency in the food system. Using the data, we are able to provide, for the first time, a comprehensive look at the national and local market positions of cooperatives and investor owned firms in the dairy and margarine products categories. Because of their regional focus, cooperatives have stronger positions in local markets than is indicated by their relatively weak national market positions. Examining aggregate national market prices for brands indicates that there are substantial premiums for many regional as well as national brands. These data are adjusted for retailer coupons but not manufacturer coupons.

When we examine selected brands in selected local markets, we find that one gains substantially more insight into a firm's brand marketing strategies. Using time series data for a particular market, it seems quite easy to estimate demand curves and elasticities at the brand level. Our refinement of the residual demand concept enables one to compare observed brand elasticities to the collusive price followship elasticities. It seems to be a useful addition to the kit of tools for measuring market power.

When looking at the brand price-share relationship for a particular brand across local markets, examples of positive, negative, and no relationship seem to surface. Our provisional theory, and we stress that our thoughts are very provisional at this time, would predict that a negative relationship is due to cost efficiencies dominating the price enhancing effect of market power. In other words the residual demand curve is more inelastic in high share markets producing a steeper marginal revenue curve but the drop in marginal costs means that the consequent profit maximizing price in large share markets is lower. Market power is still being exercised because price is greater than marginal cost, but the cost reduction more than cancels the increased market power due to more inelastic demand.

In this paper we do not test whether residual demand for a brand becomes more inelastic as the share of a particular brand increases. This needs to be done to do a complete test of the theory. When one looks across brands in a particular market, smaller share brands in fact have less, not more, elastic residual demand curves. This seems to contradict our theory and suggests that it needs to be expanded to incorporate niche effects. However, there may not be a contradiction. Different brands are designed to occupy different product niches. For example, Hood regular and Hood Light ice cream, Ben and Jerry's ice cream and Breyers ice cream each are targeted at a particular niche or market segment. The industry is segmented into strategic groups with mobility barriers between them. However, changing the price or merchandizing strategies of one of these brands does not move them to a new niche. Thus, the observed negative price-share relationship for Breyers is not due to Philip Morris positioning Breyers in a high price low share niche in one market and a low price-large share niche in another market.

A counter argument is that brands with a negative share-price relationship have inelastic demand at low rather than high share levels. This would be the case if the underlying preference structure for the product varies from one local market to the next and offsets the shift towards inelasticity as brand share increases.

The analysis of niche or strategic group effects requires that one pool several brands to capture the positioning of one brand versus another. This observation leads us to a major conclusion. The test of the market share-price relationship reported in this paper may be unduly restrictive. Here, we have examined the share-price relationship for only single brands across local markets. Large firms often market more than one brand in a market to serve different market niches and in most markets several smaller firms supply distinctive brands in particular niches. Also, the market share of a brand is not a complete measure of market structure and the prices of individual brands say little about the general price level of all brands in a market.

Work by Haller on cottage cheese and work in progress by Rogers on butter that pool all brands in the product category indicate that share-price relationships do exist across brands in more comprehensive models that control for costs and other determinants of brand price levels. For butter, Rogers does document that niche effects are important. Very small share brands do have higher prices, but price bottoms out and turns up in a quadratic fashion as share increases. This is consistent with the observed less elastic demand for some small share brands in this paper. Concerning interbrand competition, Haller demonstrates that cooperatives seem to prefer to maximize volume rather than profits on their cottage cheese sales, because there is no share-price relationship for cooperatives and their presence in a local market significantly reduces IOF branded cottage cheese prices. Moreover, both Rogers and Haller find a marginally significant positive impact of retailer concentration upon butter and cottage cheese prices.

The preliminary evidence presented here suggests that the most fruitful way to proceed may be to interview brand managers and other company executives to learn more about their marketing strategies. On this issue Ted Simmons, editor of Supermarket News reports that leading food manufacturing executives declare:

> the European model of retailing and distribution is coming to the U.S.... Each manufacturer must cut a specific deal for each account. There are no broad national or regional marketing programs. As the American retail industry keeps consolidating there will be a stronger trend toward the European model. Regional marketing is dead in the U.S. The only thing that really matters is account specific marketing (Simmons, p. 2).

Apparently this was not the case in 1989 for the brands analyzed here, otherwise diverting or centralized purchasing would have thwarted observed positive or negative relationships across local markets. Large questions remain unanswered. How, for example, do Philip Morris' executives explain the reported negative price-share relationship for Breyers Ice Cream and Parkay margarine? Are observed price differences, production, transport or distribution costs justified?

Answers to questions such as these will propel progress in marketing research. In our opinion, we are at the advent of a renaissance in applied agricultural marketing research. It will provide major new insights into the demand for food products, industrial organization models of oligopolistic food markets, and the performance of the food system. Scanner data clearly have come of age.

TABLE 5.1 Skim / Low Fat Milk: The Largest 20 Firms and All Cooperatives, 1989

Co-op Mfr Brand	Mfr Vol (1000 gals)	Brd Vol (1000 gals)	Market Share	Avg Pr. per gal	Units per gal	No of Mkts	Frequency of Rank			
							#1	#2	#3	#4
PRIVATE LABEL	1288142		63.22	2.09	1.33					
1 BORDEN INC	59109		2.90	2.37	1.46	27	6	8	5	4
BORDEN		27135	1.33	2.47	1.51	18	2	6	3	3
MEADOW Gold		22129	1.09	2.18	1.25	8	2	1	1	1
VALLEY BELL		2843	0.14	2.42	1.42	1	0	0	0	0
BORDEN VIVA		2323	0.11	2.60	2.48	3	0	0	0	1
FARMSTEAD		2272	0.11	1.73	1.00	2	0	0	1	0
LITE LINE		1801	0.09	3.47	2.55	3	1	0	0	0
KNIGHTS		274	0.01	4.02	2.15	1	0	1	0	0
2 DEAN FOODS INC	49707		2.44	2.07	1.31	10	4	0	2	2
DEANS		32447	1.59	2.23	1.46	10	4	1	1	1
FIELDCREST		15927	0.78	1.77	1.00	3	0	2	0	1
VERIFINE		1048	0.05	1.86	1.16	1	0	0	0	0
DOLLS		151	0.01	1.75	1.08	1	0	0	0	0
3 MARIGOLD FOODS INC	44716		2.19	1.98	1.15	3	0	1	1	1
KEMPS		39976	1.96	1.96	1.10	3	0	2	0	0
QUALITY CHEKD		2540	0.12	2.37	2.03	1	0	0	0	0
4 PHILIP MORRIS CO INC	33855		1.66	2.24	1.61	18	5	1	4	1
SEALTEST		16783	0.82	2.26	1.45	15	2	4	2	1
KNUDSEN		11265	0.55	2.18	1.83	2	2	0	0	0
PURITY		3479	0.17	2.54	1.42	1	1	0	0	0
LIGHT N LIVELY		2091	0.10	1.74	1.94	4	0	0	0	0
5 HILLSIDE OLD MEADOW	26669		1.31	1.93	1.42	6	2	1	0	0
HILLSIDE		26448	1.30	1.93	1.42	6	2	1	0	0
6 ANDERSON ERICKSON D	21301		1.05	1.94	1.23	1	0	0	1	0
ANDERSON ERICKSON		21301	1.05	1.94	1.23	1	0	0	1	0
7 ARABIAN INVESTMENT	18583		0.91	2.74	2.63	1	1	0	0	0
DELWOOD		18583	0.91	2.74	2.63	1	1	0	0	0
8 C AGWAY INC	17964		0.88	2.65	1.95	4	1	3	0	0

(continues)

TABLE 5.1 (continued)

#	Co-op Mfr Brand	Mfr Vol (1000 gals)	Brd Vol (1000 gals)	Market Share	Avg Pr. per gal	Units per gal	No of Mkts	#1	#2	#3	#4
	C HOOD		14389	0.71	2.46	1.79	4	1	3	0	0
	C NUFORM		3535	0.17	3.40	2.58	4	0	0	1	0
9	C DARIGOLD, INC	15784			2.20	1.66					
	C DARIGOLD		13334	0.77	2.21	1.62	2	2	0	0	0
	C DARIGLOD NU FRESH		1692	0.08	2.20	2.09	2	2	0	0	0
	C ALPINE		532	0.03	2.25	1.00	1	0	0	1	0
10	FARMLAND DRY	14451		0.71	2.41	1.74	1	0	1	0	0
11	CHIEF FRANCISCO INC D	13198		0.65	2.33	1.69	2	1	0	1	0
	TUSCAN FARMS		13198	0.65	2.33	1.69	3	0	1	1	2
12	C HILAND DRY INC	12895		0.63	2.19	1.24	3	0	0	0	3
	C HILAND		12895	0.63	2.19	1.24	3	0	0	0	3
13	REITER FOODS INC	10525		0.52	1.66	1.13	3	1	0	1	0
	REITER		10525	0.52	1.66	1.13	3	1	0	1	0
14	LEHIGH VALLEY	10277		0.50	2.19	1.77	1	0	1	0	0
	LEHIGH VALLEY		9701	0.48	2.15	1.72	1	0	1	0	1
	SEALTEST LIGHT N LIV		550	0.03	2.82	2.76	1	0	0	0	1
15	SUNNYDALE FARMS INC	10235		0.50	2.44	1.70	1	0	0	0	1
	SUNNYDALE FARMS		10235	0.50	2.44	1.70	1	0	0	0	1
16	SMITH DRY PRODS CO	9568		0.47	1.75	1.16	2	1	1	0	0
	SMITHS		9521	0.47	1.75	1.16	2	1	1	0	0
17	JOHANNA FARMS INC	9066		0.44	2.12	1.50	1	1	0	0	0
	JOHANNA FARMS		9066	0.44	2.12	1.50	1	1	0	0	1
18	SUPERIOR DRY FRESH M	8905		0.44	2.02	1.42	1	1	1	0	1
	DAIRY FRESH		8905	0.44	2.02	1.42	1	1	0	0	0
19	C PRAIRIE FARMS DRY INC	8848		0.43	2.19	1.32	4	1	1	0	1
	C PRAIRIE FARMS		8273	0.41	2.21	1.34	4	0	1	1	1
20	GARELICK FARMS INC	8727		0.43	2.64	1.75	3	1	1	0	1
	GARELICK FARMS		8636	0.42	2.63	1.72	2	1	1	0	0
27	C GOLDEN GUERNSEY DRY	7179		0.35	2.08	1.41	2	1	0	0	0
	C GOLDEN GUERNSEY		7179	0.35	2.08	1.41	2	1	0	0	0
29	C LAND O'LAKES, INC	6004		0.29	2.29	1.88	1	0	0	0	1

		Firm			0.15	2.13	1.90	1	0	0	0	0	1
33	C	LAND O'LAKES		3023	0.26	2.47	1.59	6	1	1	1	0	3
	C	FLAV-O-RICH (DAIRYMEN, INC)	5374		0.23	2.39	1.49	6	1	1	1	0	2
38	C	FLAV O RICH	4293	4703	0.21	2.08	1.37	1	1	1	0	0	0
	C	NIAGARA MILK COOP/W		4209	0.21	2.08	1.37	1	1	1	0	0	0
40	C	WENDTS	4144		0.20	2.06	1.55	1	0	0	1	0	0
	C	UPSTATE MILK COOP INC		2935	0.14	2.02	1.77	1	0	0	1	1	0
	C	SEALTEST		1208	0.06	2.14	1.00	1	1	1	0	1	1
42	C	UPSTATE	3928		0.19	2.02	1.30	2	1	1	1	0	0
	C	ROBERTS DRY CO		3598	0.18	2.06	1.34	2	1	1	0	1	1
	C	ROBERTS		330	0.02	1.55	0.78	1	1	0	0	0	1
43	C	SWEETCLOVER	3845		0.19	2.02	1.43	1	0	0	1	0	0
	C	SWISS VALLEY FARMS CO		3845	0.19	2.02	1.43	1	1	1	1	0	0
	C	SWISS VALLEY FARMS			0.15	2.06	1.24	2	1	1	1	0	0
56	C	FARM FRESH DRY INC	3031	2569	0.13	2.02	1.20	2	0	0	1	0	0
	C	FARM FRESH		204	0.01	2.66	2.00	1	0	0	0	1	0
	C	FARM FRESH HEALTH BREAK		172	0.01	1.89	1.10	1	0	0	1	0	0
	C	COUNTRY GIRL		87	0.00	1.98	1.00	1	1	1	0	1	0
63	C	SNYDERS	2383		0.12	2.09	1.40	1	0	0	0	0	1
	C	ZARDA BROTHERS DRY		2383	0.12	2.09	1.40	1	1	1	1	0	1
64	C	ZARDA	2372		0.12	2.09	1.67	1	0	0	1	0	0
	C	INTERMOUNTAIN MILK		2372	0.12	2.09	1.67	1	1	1	1	0	1
67	C	CREAM O WEBER	2141		0.11	2.25	2.15	1	0	0	0	0	0
	C	DAIRYLEA COOP INC		2141	0.05	2.25	2.15	1	0	0	0	0	0
89	C	DAIRYLEA	1011		0.05	2.61	1.35	1	0	0	0	0	1
	C	COBLE DRY PROD		1011	0.03	2.61	1.35	1	0	0	0	0	0
100	C	COBLE	640		0.03	2.11	1.69	2	0	0	0	0	2
	C	VALLEY OF VIRGINIA		640	0.02	2.11	1.69	2	0	0	0	0	1
	C	SHENANDOAH PRIDE			0.02	2.03	1.42	1	0	0	0	0	0
112	C	DAIRYMEN'S CREAMERY	467	467		2.03	1.42	1	1	0	0	0	0
	C	DAIRYMENS											

Note: The cutoff for inclusion is that a firm must have 0.5 percent of sales of skim/low fat milk in at least one local market. Thus very small local cooperatives are not included.

Source: Information Resources Inc.

TABLE 5.2 Whole Milk: The Largest 20 Firms and All Cooperatives, 1989

Co-op Mfr Brand	Mfr Vol (1000 gals)	Brd Vol (1000 gals)	Market Share	Avg Pr. per gal	Units per gal	No of Mkts	Frequency of Rank			
							#1	#2	#3	#4
PRIVATE LABEL	792439		65.70	2.25	1.32					
1 BORDEN INC	55713		4.62	2.90	1.66	27	9	8	4	4
BORDEN		43003	3.57	2.99	1.72	20	6	6	4	3
MEADOW GOLD		8918	0.74	2.47	1.51	8	3	0	1	2
FARMSTEAD		962	0.08	2.05	1.00	2	0	0	0	0
KNIGHTS		335	0.03	4.27	2.25	1	0	1	0	0
2 PHILIP MORRIS CO INC	27795		2.30	2.33	1.57	19	5	1	6	2
SALTEST		13628	1.13	2.53	1.64	17	2	3	3	4
KNUDSEN		7694	0.64	2.27	1.64	2	2	0	0	0
LIGHT N LIVELY		3847	0.32	1.47	1.01	4	0	0	1	1
PURITY		2624	0.22	2.73	1.79	1	1	0	0	0
3 DEAN FOODS CO	20724		1.72	2.50	1.48	8	4	1	1	1
DEANS		14746	1.22	2.62	1.64	8	3	2	1	1
FIELDCREST		5476	0.45	2.13	1.00	3	0	2	0	0
VERIFINE		220	0.02	2.22	1.30	1	0	0	0	1
VITA		89	0.01	2.21	2.00	1	0	0	1	0
BOWMAN		69	0.01	3.29	3.66	1	0	0	0	0
FOREST HILL		51	0.00	2.69	1.42	1	1	0	0	0
4 ARABIAN INVEST BNKG	16195		1.34	2.76	2.55	1	1	0	0	0
DELLWOOD		16195	1.34	2.76	2.55	1	1	0	0	0
5 HILLSIDE OLD MEADOW	12236		1.01	2.17	1.50	5	1	2	0	0
HILLSIDE		12201	1.01	2.17	1.50	5	1	2	1	0
6 LEHIGH VALLEY	11304		0.94	2.74	2.36	9	0	0	0	3
LEHIGH VALLEY		11216	0.93	2.74	2.36	9	0	0	1	3
7 FARMLAND DRY	10980		0.91	2.49	1.77	1	0	0	0	0
FARMLAND DAIRIES		10980	0.91	2.49	1.77	1	0	1	0	0
8 CHEF FRANCISCO INC	10641		0.88	2.37	1.53	1	0	0	0	0
TUSCAN FARMS		10641	0.88	2.37	1.53	1	0	0	1	0
9 UTOTEM INC	9770		0.81	2.38	1.50	2	1	1	0	0
ABBOTTS		8477	0.70	2.39	1.51	1	1	0	0	0

Rank		Company	Code 1	Code 2				N				
		FAIRMONT		860	0.07	2.35	1.55	1	1	0	0	0
		COUNTRY CLUB		433	0.04	2.22	1.16	1	0	0	1	0
10		WHITMAN CORP	9673									
11		PET		9519								
		SUNNYDALE FARMS INC	8715		0.80	2.20	1.50	5	0	0	1	4
		SUNNYDALE FARMS		8715	0.79	2.19	1.50	5	0	0	1	3
12	C	HILAND DRY INC	8229		0.72	2.71	1.62	1	0	0	0	1
	C	HILAND		8229	0.72	2.47	1.27	3	0	0	0	2
13		MARIGOLD FOODS INC	6000		0.68	2.47	1.27	3	0	0	1	2
		KEMPS		5998	0.68	2.27	1.32	3	0	0	0	0
14	C	FLAV-O-RICH INC	5755		0.50	2.27	1.31	6	1	0	1	0
	C	FLAV O RICH		5297	0.50	2.76	1.94	6	2	1	0	0
15		PENN DRY INC	5695		0.48	2.14	1.90	1	3	1	2	0
		PENN SUPREME		5695	0.44	2.14	1.34	1	2	1	1	0
16		CRYSTAL CREAM & BUT	5641		0.47	2.07	1.34	2	0	1	1	0
		CRYSTAL		5597	0.47	2.07	1.35	2	0	0	0	0
17		JOHANNA FARMS INC	4818		0.47	2.20	1.31	1	0	1	0	0
		JOHANNA FARMS		4745	0.46	2.20	1.57	1	0	0	0	0
18	C	AGWAY INC	4691		0.40	2.96	1.58	4	1	1	0	0
	C	HOOD		4690	0.39	2.96	2.41	4	1	1	1	0
19		CROWLEY FOODS INC	4678		0.39	2.27	2.41	1	2	1	0	0
		CROWLEY		4678	0.39	2.27	1.33	1	2	0	0	0
20		NESTLE CO	4664		0.39	2.22	1.33	4	0	0	0	0
		CARNATION		4601	0.38	2.50	1.27	4	0	2	0	0
22	C	DARIGOLD INC	3801		0.32	2.50	1.80	2	2	2	0	0
	C	DARIGOLD		3801	0.32	2.42	1.80	2	2	0	0	1
25		PRAIRIE FARMS DRY, INC	3419		0.28	2.42	1.38	4	0	0	1	0
	C	PRAIRIE FARMS		3419	0.28	2.22	1.38	4	0	1	0	0
26		FARM FRESH DRY INC	3378		0.19	2.35	1.39	2	1	1	1	1
	C	FARM FRESH		2338	0.04	1.74	1.54	2	0	0	1	0
	C	SNYDERS		459	0.03	2.13	1.00	1	1	0	0	0
	C	COUNTRY GIRL		387	0.02	1.98	1.10	1	0	0	0	0
	C	BESTYET		195			1.00	1	1	1	0	0
41	C	GOLDEN GUERNSEY	2234		0.19	2.17	1.43	2	1	1	0	0
	C	GOLDEN GUERNSEY		2234	0.19	2.17	1.43	2	0	0	1	1

(continues)

TABLE 5.2 *(continued)*

Co-op Mfr Brand	Mfr Vol (1000 gals)	Brd Vol (1000 gals)	Market Share	Avg Pr. per gal	Units per gal	No of Mkts	Frequency of Rank			
							#1	#2	#3	#4
49 C NIAGARA MILK COOP/W	1514		0.13	2.22	1.29	1	1	0	0	0
C WENDTS		1514	0.13	2.22	1.29	1	1	0	0	0
50 C COBLE DRY PROD COOP	1498		0.12	2.72	1.38	1	0	0	0	0
C COBLE		1498	0.12	2.72	1.38	1	1	0	0	0
54 C ZARDA BROTHERS DRY	1426		0.12	2.27	1.37	1	1	0	0	0
C ZARDA		1426	0.12	2.27	1.37	1	1	0	0	0
59 C ROBERTS DRY CO	1269		0.11	2.31	1.55	2	1	0	1	0
C ROBERTS		1269	0.11	2.31	1.55	2	1	0	0	1
64 C UPSTATE MILK COOP INC	1115		0.09	2.46	2.10	1	0	1	0	0
. SEALTEST		929	0.08	2.45	2.25	1	0	1	0	0
C UPSTATE		187	0.02	2.54	1.36	1	0	0	0	1
74 C LAND O'LAKES INC	847		0.07	2.50	1.98	1	0	0	0	1
C LAND O LAKES		471	0.04	2.42	2.01	1	0	0	0	1
79 C INTERMOUNTAIN MILK	816		0.07	2.49	2.33	2	0	1	0	1
C CREAM O WEBER		816	0.07	2.49	2.33	2	0	1	0	1
88 C SWISS VALLEY FARMS	668		0.06	2.64	2.18	1	0	0	0	0
C SWISS VALLEY FARMS		668	0.06	2.64	2.18	1	0	0	0	0
92 C VALLEY OF VIRGINIA	579		0.05	2.52	1.94	2	0	0	0	1
C SHENANDOAH PRIDE		579	0.05	2.52	1.94	2	0	0	0	0
135 C DAIRYMEN'S CREAMERY	56		0.00	2.43	1.37	1	0	0	1	0
C DAIRYMENS		56	0.00	2.43	1.37	1	0	0	0	1

Note: The cutoff for inclusion is that a firm must have 0.5 percent of sales of whole milk in at least one local market. Thus very small local cooperatives are not included.
Source: Information Resources Inc.

TABLE 5.3 Cottage Cheese: The Largest 20 Firms and All Cooperatives, 1989

Co-op Mfr Brand	Mfr Vol (1000 gals)	Brd Vol (1000 gals)	Market Share	Avg Pr. per gal	Units per gal	No of Mkts	Frequency of Rank			
							#1	#2	#3	#4
PRIVATE LABEL	234145		40.54	1.01	0.79					
1 PHILIP MORRIS CO	121505		21.04	1.36	0.91	40	22	7	9	1
LIGHT N LIVELY		45967	7.96	1.35	0.89	31	14	7	7	1
KNUDSEN		31368	5.43	1.43	0.97	5	3	1	1	0
BREAKSTONE		22589	3.91	1.42	0.91	33	0	6	9	6
SEALTEST		19416	3.36	1.20	0.87	29	0	8	8	3
PURITY		1168	0.20	1.27	0.92	1	1	0	0	0
KNUDSEN NICE N LIGHT		997	0.17	1.52	1.00	4	0	0	1	1
2 BORDEN INC	26424		4.57	1.22	0.86	25	6	6	3	4
BORDEN		10526	1.82	1.36	0.95	16	3	3	2	4
BORDEN VIVA		7493	1.30	1.07	0.79	7	2	1	1	0
LITE LINE		4292	0.74	1.34	0.79	12	0	3	0	2
MEADOW GOLD		3625	0.63	1.03	0.79	7	0	1	2	0
GREAT SCOTT		239	0.04	0.89	0.67	1	0	0	0	0
3 C AGWAY INC	21813		3.78	1.25	0.99	6	5	0	0	0
C HOOD		21813	3.78	1.25	0.99	6	5	0	1	1
4 QLTY CHEKD DRY PRO	16333		2.83	0.99	0.77	14	1	2	1	2
KEMPS SLIM TRIM		13295	2.30	1.01	0.77	13	1	2	0	3
QUALITY CHEKD		2363	0.41	0.90	0.73	4	0	0	0	0
BAY VIEW FARMS		479	0.08	0.99	0.72	1	0	0	1	0
CURLYS		101	0.02	0.79	0.97	1	0	0	0	0
5 FRIENDSHIP FOOD PROD	13038		2.26	1.48	1.08	5	0	4	1	0
FRIENDLY FARMER		10289	1.78	1.43	0.94	4	1	1	1	0
FRIENDSHIP		2749	0.48	1.69	1.62	4	0	0	1	0
6 DEAN FOODS CO	11694		2.02	1.33	0.94	10	0	5	0	1
DEANS		10809	1.87	1.36	0.95	8	1	2	2	0
FIELDCREST		664	0.11	0.79	0.67	2	0	0	0	1

(continues)

TABLE 5.3 (continued)

Co-op Mfr Brand	Mfr Vol (1000 gals)	Brd Vol (1000 gals)	Market Share	Avg Pr. per gal	Units per gal	No of Mkts	#1	#2	#3	#4
							| Frequency of Rank			
BOWMAN		90	0.02	1.40	0.88	1	0	0	0	0
ALL JERSEY		53	0.01	0.88	0.67	1	0	0	0	0
VERIFINE		45	0.01	1.30	0.92	1	0	0	0	0
7 CROWLEY FOODS, INC	10163		1.76	1.20	0.93	12	0	5	3	2
CROWLEY		5417	0.94	1.11	0.85	6	0	1	3	2
AXELROD		4747	0.82	1.31	1.02	7	0	0	1	3
8 ANDERSON ERICKSON	6655		1.15	0.98	0.84	2	0	0	0	0
ANDERSON ERICKSON		6655	1.15	0.98	0.84	2	0	0	1	0
9 C DARIGOLD, INC	6514		1.13	0.93	0.82	2	2	0	0	0
C DARIGOLD		6514	1.13	0.93	0.82	2	2	0	0	0
10 H J HEINZ CO	6263		1.08	1.20	0.96	23	0	4	1	8
WEIGHT WATCHERS		6263	1.08	1.20	0.96	23	0	2	3	4
11 MARIGOLD FOODS INC	5551		0.96	1.12	0.75	4	0	1	0	0
QUALITY CHEKD		2867	0.50	1.07	0.67	3	0	0	1	0
KEMPS		1354	0.23	1.25	0.99	2	0	1	0	0
KEMPS LITE		1330	0.23	1.10	0.67	3	0	0	1	0
12 C PRAIRIE FARMS DRY, INC	4713		0.82	1.07	0.76	4	0	0	1	2
C PRAIRIE FARMS		4504	0.78	1.09	0.76	4	0	1	0	1
13 GENERAL MILLS	4542		0.79	1.63	0.87	6	1	2	0	0
MICHIGAN		4542	0.79	1.63	0.87	6	2	0	1	0
14 C INTERMOUNTAIN MILK	4261		0.74	1.00	0.82	3	1	1	0	1
C CREAM O WEBER		4261	0.74	1.00	0.82	3	1	0	1	1
15 OLD HOME FOODS INC	4019		0.70	1.75	0.92	1	1	0	0	0
OLD HOME		3367	0.58	1.75	0.93	1	1	0	0	0
SLENDRELLA		653	0.11	1.73	0.84	1	0	0	1	0
16 NORDICA INTL INC	3949		0.68	1.24	0.82	2	0	1	0	0

No.		Firm / brand	Firm total	Brand										
		NORDICA		3944	0.68	1.24	0.82	2	0	1	0	0	0	0
17	C	GOLDEN GUERNSEY DRY	3291		0.57	1.23	0.85	1	1	1	1	0	0	0
18	C	GOLDEN GUERNSEY		3291	0.57	1.23	0.85	1	1	1	0	0	0	2
	C	CABOT FRMRS' COOP	2982		0.52	0.88	0.98	5	0	0	0	3	3	2
19	C	CABOT		2982	0.52	0.88	0.98	5	0	0	0	2	2	1
		NESTLE CO	2745		0.48	0.97	0.85	3	0	1	1	1	1	0
20		CARNATION		2745	0.48	0.97	0.85	3	0	1	1	1	1	0
		REITER FOODS INC	2736		0.47	0.86	0.73	3	1	0	0	1	1	2
21		REITER		2736	0.47	0.86	0.73	2	1	1	1	0	0	0
	C	FARM FRESH DRY INC	2640		0.46	0.94	0.75	2	1	0	1	1	1	1
	C	FARM FRESH		2123	0.37	0.97	0.76	2	0	1	1	0	0	0
	C	COUNTRY GIRL		448	0.08	0.81	0.67	2	1	0	1	1	1	1
	C	FARM FRESH HEALTH		69	0.01	1.07	1.00	1	0	0	0	0	0	0
24	C	BISON FOODS CO	2329		0.40	1.19	0.93	1	1	1	1	0	0	0
	C	BISON		2329	0.40	1.19	0.93	1	1	1	1	1	1	1
31	C	ROBERTS DRY CO	1488		0.26	0.94	0.77	2	1	0	1	0	0	0
	C	ROBERTS		1488	0.26	0.94	0.77	2	1	1	1	1	1	1
34	C	SWISS VALLEY FARMS CO	1161		0.20	1.14	0.81	1	0	0	0	0	0	0
	C	SWISS VALLEY FARMS		1161	0.20	1.14	0.81	1	1	0	1	0	0	1
35	C	FLAV-O-RICH INC	1127		0.20	1.13	0.93	3	0	0	0	0	0	0
	C	FLAV O RICH		1127	0.20	1.13	0.93	3	0	0	0	1	1	1
38	C	ZARDA BROTHERS DRY	823		0.14	0.99	0.74	3	1	1	1	0	0	0
	C	ZARDA		823	0.14	0.99	0.74	2	1	1	1	0	0	0
40	C	LAND O'LAKES, INC	804		0.14	1.23	0.76	2	1	1	1	0	0	0
	C	LAND O LAKES		511	0.09	1.17	0.72	1	1	0	1	1	1	1
47	C	DAIRYLEA COOP INC	521		0.09	1.01	1.02	1	0	0	0	0	0	0
	C	DAIRYLEA		521	0.09	1.01	1.02	2	0	0	0	0	0	0
53	C	VALLEY OF VIRGINIA	399		0.07	0.94	0.79	2	1	0	1	0	1	0
	C	SHENANDOAH PRIDE		399	0.07	0.94	0.79	2	0	0	0	0	0	1

Note: The cutoff for inclusion is that a firm must have 0.5 percent of sales of cottage cheese in at least one local market. Thus very small local cooperatives are not included.
Source: Information Resources Inc.

TABLE 5.4 Butter: The Largest 20 Firms and All Cooperatives, 1989

Co-op Mfr Brand	Mfr Vol (1000 gals)	Brd Vol (1000 gals)	Market Share	Avg Pr. per gal	Units per gal	No of Mkts	Frequency of Rank			
							#1	#2	#3	#4
PRIVATE LABEL	137534		44.25	1.82	1.05					
1 C LAND O'LAKES INC	97695		31.43	2.11	1.08	51	38	8	3	1
C LAND O LAKES		97547	31.38	2.11	1.08	51	39	7	2	2
C AYRSHIRE		81	0.03	2.20	0.50	1	0	0	0	0
2 PHILIP MORRIS CO INC	13753		4.42	2.45	1.96	15	0	8	6	1
BREAKSTONE		12992	4.18	2.44	1.97	12	0	8	3	1
KNUDSEN		704	0.23	2.68	1.87	2	0	0	0	2
PURITY		58	0.02	2.54	1.00	1	0	1	0	0
3 BORDEN INC	11399		3.67	2.10	1.02	17	1	3	7	3
KELLERS		5403	1.74	1.99	1.00	5	1	1	1	2
HOTEL BAR		4748	1.53	2.26	1.04	1	0	0	0	2
MEADOW GOLD		772	0.25	1.80	1.01	5	0	0	1	2
BORDEN		248	0.08	2.60	1.06	6	0	1	2	0
COUNTRY STORE		230	0.07	1.86	1.00	4	0	0	1	2
4 C CHALLENGE DRY PRODS	10918		3.51	2.37	1.22	8	4	1	1	1
C CHALLENGE		9480	3.05	2.36	1.25	8	4	1	1	1
C DANISH		1427	0.46	2.43	1.05	2	0	0	2	0
5 LOV-IT CREAMERY INC	4605		1.48	1.37	0.99	1	0	1	0	0
BUTTER UP		3027	0.97	1.26	1.00	1	0	1	0	0
LOV IT		1335	0.43	1.56	0.96	1	0	0	0	1
6 CRYSTAL FOODS INC	4254		1.37	1.69	0.98	11	1	5	1	3
CRYSTAL FARMS		4254	1.37	1.69	0.98	11	1	5	1	2
7 C DARIGOLD INC	2960		0.95	1.96	1.19	5	2	0	0	2
C DARIGOLD		2364	0.76	2.06	1.24	4	2	0	0	0
C MED O DEW		511	0.16	1.46	1.00	1	0	0	1	0
8 BEATRICE FOODS CO	2863		0.92	1.81	1.05	8	0	4	0	3
SWIFT PREMIUM		1974	0.64	1.74	1.00	4	0	3	1	1
SUGAR CREEK		371	0.12	1.94	1.04	2	0	0	1	1

#	C	Name	Vol A	Vol B				n				
9	C	ASSC MILK PRDUC	2302		0.09	2.09	1.00	1	0	1	0	0
		BREDAN		265	0.04	1.97	2.00	3	0	0	0	3
		SWIFTS BROOKFIELD		113	0.02	1.82	1.00	4	1	0	3	1
		SUMNERS		48	0.74	1.93	1.06	1	0	0	1	1
	C	SOMMER MAID		1516	0.49	1.95	1.00	2	0	0	1	0
	C	STATE BRAND		458	0.15	1.81	1.28	1	0	0	2	0
	C	COUNTRY MAID		328	0.11	2.04	1.00	5	0	2	0	0
10	C	MID-AMERICA DAIRYMEN	1645	1286	0.53	1.82	1.09	4	0	1	1	1
	C	MID AMERICA FARMS		250	0.41	1.81	1.09	1	0	0	1	0
	C	RECIPE BOOK		87	0.08	1.67	1.00	1	0	1	0	1
	C	COLORADO			0.03	2.34	1.00	1	0	0	0	0
11	C	CABOT FRMRS' COOP	1321	1321	0.42	2.02	1.00	4	0	1	1	1
	C	CABOT			0.42	2.02	1.00	4	0	3	1	0
12	C	TILLAMOOK CNTY CR	1213	1213	0.39	2.00	1.00	3	0	3	0	0
	C	TILLAMOOK			0.39	2.00	1.00	3	0	2	0	0
13	C	PRAIRIE FARMS DRY INC	954	954	0.31	1.81	1.02	1	0	2	0	0
	C	PRAIRIE FARMS			0.31	1.81	1.00	1	0	1	0	0
14		QLTY CHEKD DRY PRO	918	730	0.30	1.87	1.00	7	1	1	1	0
		KEMPS SLIM TRIM		112	0.23	1.87	1.00	2	1	0	0	0
		SINTONS		71	0.04	1.73	1.00	1	1	1	1	3
		CURLYS			0.02	2.12	1.00	3	1	0	0	2
15		UNTD DAIRYMEN OF AZ	887	887	0.29	1.99	1.00	1	0	0	0	1
		SEAL OF ARIZONA			0.29	1.99	1.00	1	1	0	0	0
16	C	CACHE VALLEY DRY ASSN	832	832	0.27	1.86	1.00	3	1	0	0	0
	C	CACHE VALLEY			0.27	1.86	1.00	3	1	0	0	0
17		LOUBAT-L. FRANK, INC	816	816	0.26	2.23	1.16	1	1	0	0	0
		AMERICAN BEAUTY			0.26	2.23	1.16	1	1	0	0	0
18		LEVEL VALLEY DRY CO	741	393	0.24	1.65	1.00	2	0	0	1	0
		VALLEY MAID		349	0.13	1.66	1.00	1	0	0	1	0
		LEVEL VALLEY			0.11	1.64	1.00	1	0	1	0	0
19		CRYSTAL CREAM & BUT	715	715	0.23	1.72	1.00	2	0	1	1	0
		CRYSTAL			0.23	2.01	1.00	2	0	0	1	0
20		ORRVILLE MILK DIV	685		0.22	1.73	1.00	1	0	0	0	0

(continues)

TABLE 5.4 (continued)

Co-op Mfr Brand	Mfr Vol (1000 gals)	Brd Vol (1000 gals)	Market Share	Avg Pr. per gal	Units per gal	No of Mkts	Frequency of Rank #1	#2	#3	#4
COTTAGE		685	0.22	1.73	1.00	1	0	1	0	0
21 C BONGARDS CREAMERIES	609		0.20	1.55	1.00	1	0	0	1	0
C BONGARDS		609	0.20	1.55	1.00	1	0	0	1	0
22 C GOLDEN GUERNSEY DRY	580		0.19	1.71	1.00	1	0	0	1	0
C GOLDEN GUERNSEY		580	0.19	1.71	1.00	1	0	0	1	0
23 C CALIF COOP CREAMERY	497		0.16	1.78	1.00	1	1	1	1	0
C CALIFORNIA GOLD		497	0.16	1.78	1.00	1	1	1	1	0
27 C INTERMOUNTAIN MILK	385		0.12	2.08	1.00	2	0	1	1	0
C CREAM O WEBER		337	0.11	2.08	1.00	2	0	1	1	0
C CRYSTAL		48	0.02	1.72	1.00	2	0	0	0	0
31 C MICHIGAN MILK PRDUC	254		0.08	1.78	0.92	2	0	1	0	1
C RAMUS		146	0.05	1.74	1.00	1	0	1	0	0
C MICHIGAN		109	0.03	1.83	0.82	2	0	0	1	1
41 C O-AT-KA MILK PRODS COOP	125		0.04	1.85	1.00	1	0	0	0	1
C GOLD COW		125	0.04	1.85	1.00	1	0	1	1	1
42 C FARM FRESH DRY INC	93		0.03	2.48	1.00	2	0	0	0	1
C FARM FRESH		93	0.03	2.48	1.00	2	0	1	0	1
45 C ROBERTS DRY CO	88		0.03	2.17	1.00	1	1	0	0	0
C ROBERTS		88	0.03	2.17	1.00	1	1	0	0	0
58 C ZARDA BROTHERS DRY INC	38		0.01	2.14	1.00	1	0	0	1	0
C ZARDA		38	0.01	2.14	1.00	1	0	0	1	0
63 C DAIRYMEN'S CREAMERY	19		0.01	2.06	1.00	1	0	0	0	0
C DAIRYMENS		19	0.01	2.06	1.00	1	0	0	0	0

Note: The cutoff for inclusion is that a firm must have 0.5 percent of sales of butter in at least one local market. Thus very small local cooperatives are not included.
Source: Information Resource Inc.

TABLE 5.5 Margarine and Spreads: All Firms, 1989

Co-op Mfr Brand	Mfr Vol (1000 gals)	Brd Vol (1000 gals)	Market Share	Avg Pr. per gal	Units per gal	No of Mkts	Frequency of Rank			
							#1	#2	#3	#4
PRIVATE LABEL	268337		16.03	0.49	0.84					
1 UNILEVER	411802		24.59	0.88	0.84	51	25	16	10	0
SHEDDS COUNTRY		172171	10.28	0.85	0.66	51	1	16	13	5
IMPERIAL		109006	6.51	0.69	0.95	51	6	5	1	5
PROMISE		78933	4.71	1.34	0.98	51	1	0	2	7
SHEDDS		45537	2.72	0.61	0.99	51	0	0	0	0
KRONA		1651	0.10	1.16	1.00	50	0	0	0	0
ELGIN		1239	0.07	0.44	1.00	4	0	0	0	0
IMPERIAL ALA MODE		710	0.04	1.13	0.77	4	0	0	0	0
HOTEL		645	0.04	0.68	1.00	3	0	0	0	0
SPRING VALLEY		193	0.01	0.57	1.00	1	0	0	0	0
TASTEE SPREAD		184	0.01	0.55	0.41	1	0	0	0	0
BLUE SEAL		41	0.00	0.47	1.00	1	0	0	0	0
2 RJR NABISCO INC/RJR	388555		23.21	0.97	0.90	51	19	21	10	1
BLUE BONNET		192002	11.47	0.60	0.84	51	10	6	9	9
FLEISCHMANNS		191607	11.44	1.32	0.96	51	10	10	14	12
BLUE BONNET BETTER		4947	0.30	1.34	1.00	9	0	0	0	0
3 PHILIP MORRIS CO INC	286186		17.09	0.80	0.90	51	7	12	26	3
PARKAY		221528	13.23	0.79	1.00	51	20	13	8	4
PARKAY LIGHT		32046	1.91	0.70	0.41	51	0	0	0	0
KRAFT TOUCH OF BTR		23073	1.38	0.81	0.62	28	0	0	0	0
CHIFFON		7311	0.44	1.26	1.00	14	0	0	0	0
KRAFT MIRACLE WHIP		1486	0.09	1.35	1.00	2	0	0	0	0
MRS TUCKER		580	0.03	0.62	0.36	1	0	0	0	0
4 CNTRL SOYA CO INC	119940		7.16	1.21	0.98	51	0	1	1	32

(continues)

TABLE 5.5 (continued)

Co-op Mfr Brand	Mfr Vol (1000 gals)	Brd Vol (1000 gals)	Market Share	Avg Pr. per gal	Units per gal	No of Mkts	#1	#2	#3	#4
I CANT BELIEVE ITS N BTR		95267	5.69	1.37	1.00	51	0	1	3	2
MRS FILBERTS	85602	24666	1.47	0.62	0.90	15	0	0	0	1
5 C LAND O'LAKES, INC		58069	5.11	1.13	1.00	43	2	1	3	7
C LAND O LAKES		27533	3.47	0.86	1.00	39	2	1	1	2
C COUNTRY MORNING BLEND			1.64	1.70	1.00	41	0	0	0	0
6 C PC INTERNATIONAL	41018		2.45	1.13	0.94	51	0	0	0	1
MAZOLA		33175	1.98	1.16	0.94	50	0	0	0	0
NUCOA		5626	0.34	0.94	1.00	8	0	0	0	0
MAZOLA LIGHT		2217	0.13	1.25	0.78	1	0	0	0	0
7 H J HEINZ CO	21915		1.31	0.97	1.00	51	0	0	0	0
WEIGHT WATCHERS		21915	1.31	0.97	1.00	51	0	0	1	3
8 BORDEN INC	11405		0.68	0.77	0.86	9	0	0	0	3
GOLD & SOFT		9602	0.57	0.79	0.92	8	0	0	0	3
KELLERS		1130	0.07	0.67	0.68	1	0	0	0	0
GREGGS		338	0.02	0.65	0.26	1	0	0	0	0
9 PVO INTL INC	8850		0.53	1.05	1.00	7	0	0	0	2
SAFFOLA		8765	0.52	1.05	1.00	7	0	0	0	1
10 MIAMI MARGARINE CO	5243		0.31	0.54	0.97	6	0	0	0	1
NUMAID		3591	0.21	0.61	0.95	3	1	0	0	0
ROYAL SCOT		1470	0.09	0.38	1.02	4	0	0	0	0
11 DEAN FOODS CO	3346		0.20	0.36	0.59	3	0	0	0	0
DEW FRESH		3204	0.19	0.36	0.57	3	0	0	0	0
12 CRYSTAL FOODS INC	2439		0.15	0.64	0.80	3	0	0	0	0
CRYSTAL FARMS		2439	0.15	0.64	0.80	3	0	0	0	0

#	Name									
13	SUNNYLAND REFINING	2243		0.13	0.46	1.18	2	0	0	0
	SUNNYLAND		2214	0.13	0.45	1.19	2	0	0	0
14	WHITMAN CORP	2027		0.12	1.29	1.00	2	0	0	0
	HOLLYWOOD		2026	0.12	1.29	1.00	2	0	0	0
15	C CHALLENGE DRY PRO	1256		0.07	1.41	1.00	4	0	0	0
	C CHALLENGE DAIRY		1256	0.07	1.41	1.00	4	0	0	0
16	WILSEY FOODS INC	1224		0.07	0.70	0.97	4	0	0	0
	GOLD N SWEET		770	0.05	0.75	1.00	3	0	0	0
	TABLE MAID		446	0.03	0.60	0.91	1	0	0	0
17	QUAKER OATS CO	836		0.05	0.58	0.51	2	0	0	0
	VELVET SPREAD		815	0.05	0.58	0.50	2	0	0	0
18	HOLSUM FOODS	667		0.04	0.36	0.94	2	0	0	0
	CROWN		549	0.03	0.34	1.00	1	0	0	0
	TASTEE GOLD		64	0.00	0.44	0.33	1	0	0	0
19	OSCEOLA FOODS, INC	663		0.04	0.45	1.30	3	0	0	0
	DAIRY PRIDE		243	0.01	0.43	1.00	1	0	0	0
	RIVERVIEW		148	0.01	0.38	1.00	1	0	0	0
20	C ASSC MILK PRDUC	210		0.01	0.57	1.00	1	0	0	0
	C SOMMER MAID		210	0.01	0.57	1.00	1	0	0	0
21	HESS BROS FARMS INC	203		0.01	0.72	0.99	1	0	0	0
	HESS FOODS		203	0.01	0.72	0.99	1	0	0	0
22	C PRAIRIE FARMS DRY, INC	136		0.01	0.43	1.00	1	0	0	0
	C PRAIRIE FARMS		136	0.01	0.43	1.00	1	0	0	0

Note: The cutoff for inclusion is that a firm must have 0.5 percent of sales of margarine in at least one local market. Thus very small local cooperatives are not included.

Source: Information Resource Inc.

TABLE 5.6 Ice Cream: The Largest 20 Firms and All Cooperatives, 1989

Co-op Mfr	Brand	Mfr Vol (1000 gals)	Brd Vol (1000 gals)	Market Share	Avg Pr. per gal	Units per gal	No of Mkts	#1	#2	#3	#4
	PRIVATE LABEL	324006		40.78	1.88	0.97					
1	PHILIP MORRIS CO INC	122087		15.37	3.20	1.11	46	18	9	6	5
	BREYERS		72479	9.12	3.46	1.03	45	12	9	10	5
	SEALTEST		41755	5.26	2.37	1.03	29	4	11	2	3
	FRUSEN GLADJE		3073	0.39	8.30	4.00	11	0	0	0	0
	PURITY		1094	0.14	3.12	1.09	1	1	0	0	0
	KNUDSEN		1056	0.13	3.46	1.28	2	0	0	0	0
	LIGHT N LIVELY		1045	0.13	2.66	1.00	3	0	0	0	0
2	BORDEN INC	39420		4.96	2.23	1.00	36	1	17	4	7
	BORDEN		21340	2.69	2.11	0.99	25	1	2	8	5
	MEADOW GOLD		6189	0.78	2.17	0.95	10	0	3	1	1
	LADY BORDEN		3023	0.38	3.20	0.90	11	0	0	0	2
	BORDEN OLD FASH		2539	0.32	2.66	1.28	10	0	0	1	1
	GLACIER CLUB		2256	0.28	1.58	1.00	5	0	0	1	0
	FARMSTEAD		1763	0.22	1.38	1.00	5	0	0	0	1
	BORDEN EAGLE		1038	0.13	3.69	1.22	7	0	0	1	0
	MEADOW GOLD SUPR		637	0.08	2.44	1.10	2	0	0	0	1
	ROYAL DANISH		228	0.03	1.69	1.00	2	0	0	0	0
3	DREYER'S GRAND ICE	38369		4.83	4.38	1.33	22	6	5	1	4
	DREYERS EDYS		25445	3.20	4.37	1.35	22	5	4	2	3
	DREYERS EDYS GR LT		12924	1.63	4.40	1.28	21	0	3	2	2
4	BLUE BELL CREAMER	30040		3.78	3.36	1.17	5	3	1	0	0
	BLUE BELL		19587	2.47	3.33	1.02	5	3	1	0	0
	BLUE BELL SUPREME		10453	1.32	3.43	1.46	5	0	3	0	1
5	MARIGOLD FOODS INC	16955		2.13	1.81	0.66	8	1	0	1	1
	KEMPS		14870	1.87	1.86	0.62	8	1	0	1	0
	VALUE PAK		1820	0.23	1.42	0.96	4	0	0	0	0

#	Name										
6	C AGWAY INC	12974		1.63	2.06	1.01	7	2	2	1	0
	C HOOD		10829	1.36	2.04	1.01	5	3	0	1	1
	C HOOD LIGHT		1820	0.23	2.24	1.00	7	0	0	0	1
	C NUFORM		324	0.04	1.73	1.00	3	0	0	0	0
7	GRANDMET	10642		1.34	9.06	4.00	43	0	0	2	0
	HAAGEN DAZS		10642	1.34	9.06	4.00	43	0	1	2	5
8	MAYFIELD DRY FARMS	9793		1.23	2.30	1.06	2	2	0	0	1
	MAYFIELD DAIRY		9793	1.23	2.30	1.06	2	1	0	0	0
9	QLTY CHEKD DRY PROD	8515		1.07	2.21	0.97	11	2	0	2	0
	BABCOCK		3435	0.43	2.09	1.00	3	1	0	0	0
	RUGGLES		1687	0.21	3.00	1.00	3	1	0	0	0
	QUALITY CHEKD		820	0.10	2.00	1.00	4	0	0	0	0
	BAY VIEW FARMS		623	0.08	1.83	1.00	1	0	0	0	0
	FESTIVAL		566	0.07	1.57	1.00	1	0	0	0	0
	3 RIVERS		273	0.03	1.65	0.40	1	0	0	0	0
	WHALE OF A PAIL		174	0.02	1.59	0.40	1	0	0	0	0
	WATTS HARDY		122	0.02	1.47	0.76	1	0	0	0	0
	GILT EDGE FARMS		116	0.01	1.05	1.00	1	0	0	0	0
	MOVENPICK		115	0.01	4.44	1.31	1	0	0	0	0
	SUNNYSIDE FARMS		112	0.01	2.12	1.00	1	0	0	0	0
	SINTONS		101	0.01	1.90	0.72	1	0	0	0	0
	NUT TREE		68	0.01	3.51	1.00	1	0	0	0	0
10	TURKEY HILL DRY INC	8279		1.04	2.56	1.00	2	0	1	0	0
	TURKEY HILL		8279	1.04	2.56	1.00	2	0	1	0	0
11	STEVE'S HOMEMADE	6984		0.88	3.78	1.40	4	0	1	1	0
	DOLLY MADISON		3878	0.49	2.62	1.00	2	0	0	0	0
	T & W		1853	0.23	4.07	1.18	1	0	0	0	0
	STEVES		826	0.10	8.68	4.00	1	0	0	0	0
	SWENSONS		427	0.05	3.64	1.00	1	0	0	0	0
12	WHITMAN CORP	6789		0.85	2.37	1.02	6	0	0	1	3
	PET		5836	0.73	2.48	1.04	5	0	0	1	1

(continues)

TABLE 5.6 (continued)

Co-op Mfr Brand	Mfr Vol (1000 gals)	Brd Vol (1000 gals)	Market Share	Avg Pr. per gal	Units per gal	No of Mkts	#1	#2	#3	#4
COUNTRY LOVES		662	0.08	1.64	1.00	1	0	0	0	0
13 WELLS DRY INC	6385		0.80	2.08	0.88	9	1	0	3	1
BLUEBUNNY		6026	0.76	2.07	0.87	9	1	1	2	1
COUNTRY RICH		219	0.03	1.59	1.00	2	0	0	0	0
WELLS TRADITIONAL		111	0.01	3.57	1.00	1	0	0	0	2
14 C FLAV-O-RICH INC	6145		0.77	2.27	0.99	8	0	3	3	1
C FLAV O RICH		3845	0.48	2.10	1.02	8	0	0	0	1
C RICH & CREAMY		1481	0.19	3.00	1.00	7	0	0	1	0
C DAIRY CHARM		471	0.06	1.65	0.89	1	0	0	0	0
C CREAMI RICH		192	0.02	2.12	0.50	1	0	0	0	0
C DIXIELAND		156	0.02	1.52	1.00	2	0	0	0	1
15 DEAN FOODS CO	5674		0.71	2.15	1.03	5	0	2	1	0
FIELDCREST		2583	0.33	1.55	0.95	3	0	1	0	0
DEANS		2080	0.26	2.86	1.27	3	0	0	0	0
VERIFINE		515	0.06	1.98	0.65	1	0	0	0	0
COUNTRY CHARM		394	0.05	2.72	1.00	2	0	0	0	0
FAMILY PAK		90	0.01	1.48	0.40	1	0	0	0	0
16 BEN & JERRY'S HMDE	5234		0.66	8.86	4.00	20	0	0	0	2
BEN & JERRYS		5234	0.66	8.86	4.00	20	0	0	0	0
17 C PRAIRIE FARMS DRY, INC	4944		0.62	2.18	0.89	3	1	1	0	1
C PRAIRIE FARMS		3529	0.44	2.08	0.85	3	1	1	0	0
C OLD RECIPE		969	0.12	2.86	1.00	3	0	0	0	0
C FAMILY PAK		351	0.04	1.46	0.91	1	0	0	0	0
C PRAIRIE FARM SUPREME		96	0.01	1.60	1.00	1	0	0	0	0
18 UNTD DRY FARMERS INC	4604		0.58	3.73	1.21	6	1	0	1	2
HOMEMADE		4516	0.57	3.76	1.21	6	1	1	1	2
19 HENDRIE'S, INC	4432		0.56	2.16	1.04	4	0	0	3	1

#		Firm / Brand	Firm	Brand								
		HENDRIES	4432		0.56	2.16	1.04	4	0	0	1	1
20	C	DARIGOLD, INC	4090		0.51	2.04	1.00	2	1	1	1	0
	C	DARIGOLD		4081	0.51	2.04	0.99	2	2	0	0	0
22	C	INTERMOUNTAIN MILK	3662		0.46	1.99	0.87	3	2	0	0	0
	C	CREAM O WEBER		3417	0.43	2.00	0.86	3	2	0	0	0
	C	CRYSTAL		246	0.03	1.81	1.00	2	0	0	0	0
30	C	LAND O'LAKES, INC	2699		0.34	1.91	0.64	1	0	1	0	0
	C	LAND O LAKES		2041	0.26	1.77	0.60	1	0	1	0	0
	C	COUNTRY CREAMERY		131	0.02	3.35	1.00	1	2	0	0	0
38	C	FARM FRESH DRY INC	1882		0.24	1.90	1.00	2	2	0	0	0
	C	FARM FRESH		1812	0.23	1.87	0.99	2	0	0	0	0
	C	GOODTIME		29	0.00	1.58	1.00	1	0	0	0	0
	C	COUNTRY GIRL		20	0.00	1.59	1.00	1	0	0	0	0
42	C	HILAND DRY INC	1762		0.22	2.38	0.98	2	0	0	0	0
	C	HILAND		1584	0.20	2.40	0.95	2	0	0	0	0
48	C	MORNING GLORY FARMS	1415		0.18	2.00	0.67	1	0	0	0	0
	C	MORNING GLORY		1385	0.17	1.97	0.65	1	1	0	0	0
52	C	ZARDA BROTHERS DRY	1201		0.15	2.06	0.95	2	1	0	0	0
	C	ZARDA		1143	0.14	2.07	0.96	2	1	0	0	0
	C	COUNTRY FARM		38	0.00	1.95	0.40	1	0	0	0	0
63	C	COBLE DRY PROD. COOP	820		0.10	1.66	0.40	1	0	0	0	0
	C	COBLE		820	0.10	1.66	0.40	1	0	0	0	0
86	C	ROBERTS DRY CO	405		0.05	2.13	0.97	2	0	0	1	0
	C	ROBERTS		385	0.05	2.13	1.00	2	0	1	1	0
115	C	TILLAMOOK CNTY CR	97		0.01	3.77	1.00	1	0	1	0	0
	C	TILLAMOOK		97	0.01	3.77	1.00	1	0	0	0	0
127	C	DAIRYMEN'S CREAM	20		0.00	3.70	1.05	1	0	0	0	0
	C	DAIRYMENS		20	0.00	3.70	1.05	1	0	0	0	0

Note: The cutoff for inclusion is that a firm must have 0.5 percent of sales of ice cream in at least one local market. Thus very small local cooperatives are not included.

Source: Information Resources Inc.

TABLE 5.7 Market Positions of Cooperatives in Six Product Categories: 1989.

Category	National Rank of Coops in top 20	No. of Coops in Category	No. of Local Markets	Frequencies of Share Rank for all Cooperatives				
				1	2	3	4	5 or >
Low fat/skim milk	8, 9, 12, 19	18	36	8	11	1	10	6
Whole milk	12, 14, 18	16	36	8	9	5	7	7
Cottage cheese	3, 9, 12, 14, 17, 18	15	37	13	3	7	9	5
Butter	1, 4, 7, 9, 10, 11, 12, 13, 16	19	97	47	20	13	9	8
Ice Cream	6, 14, 17, 20	14	36	9	9	4	3	11
Market position totals for five dairy product categories	-	-	242	85	53	33	45	74
Percent Penetration[a]				33.3	20.8	12.9	17.6	-

[a] Defined as the number of coops in rank divided by the total number of positions available (5 categories X 51 local markets = 255 positions) expressed as a percent.

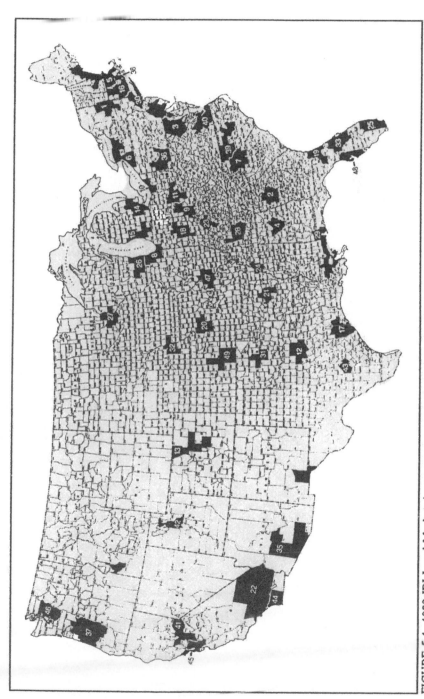

FIGURE 5.1 1989 IRI Local Market Areas

Source: Progressive Grocer's 1990 Market Scope, Maclean Hunter Media, Inc., Stamford, CT.

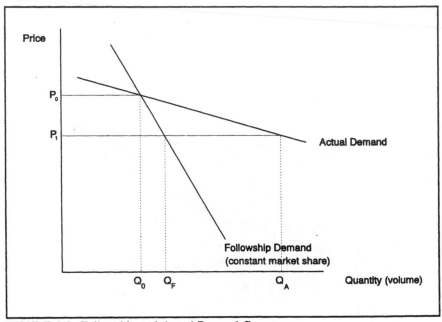

FIGURE 5.2 Followship and Actual Demand Curves

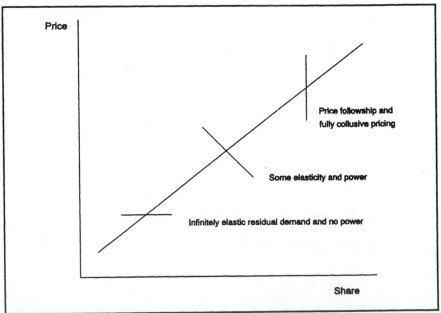

FIGURE 5.3   The Relationship Between Price and Brand Share for Branded Food Products

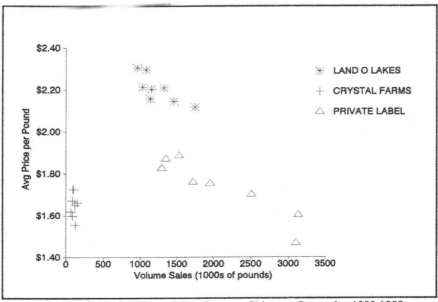

FIGURE 5.4   Price and Volume Sales, Butter - Chicago: Quarterly, 1988-1989

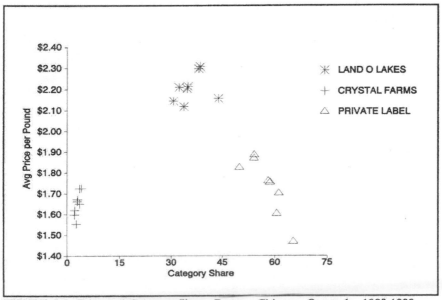

FIGURE 5.5   Price and Category Share, Butter - Chicago: Quarterly, 1988-1989

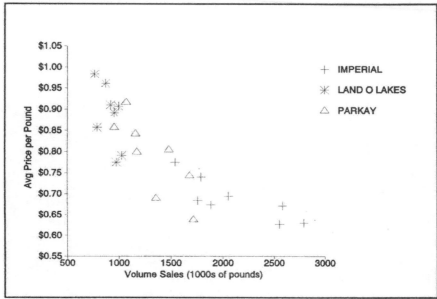

FIGURE 5.6 Price and Volume Sales, Margarine - Chicago: Quarterly 1988 - 1989

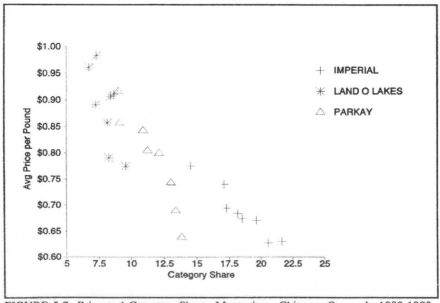

FIGURE 5.7 Price and Category Share: Margarine - Chicago: Quarterly 1988-1989

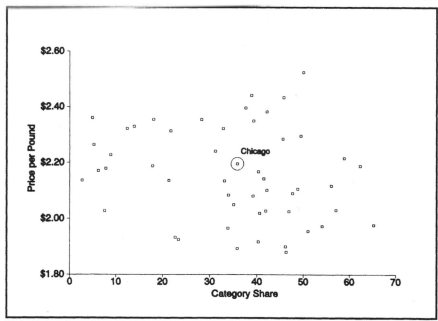

FIGURE 5.8 Price and Share, Butter - Land O'Lakes: 1989 Annual Data

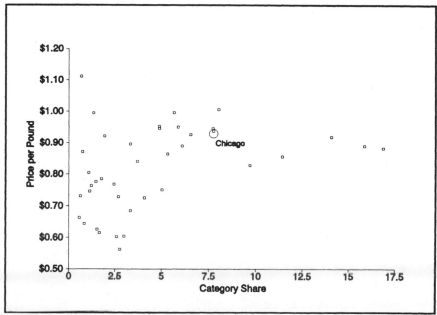

FIGURE 5.9 Price and Category Share, Land O'Lakes Margarine: Annual 1989

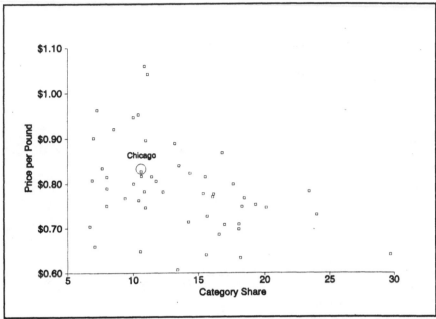

FIGURE 5.10 Price and Category Share, Parkay Margarine-Philip Morris: Annual 1989

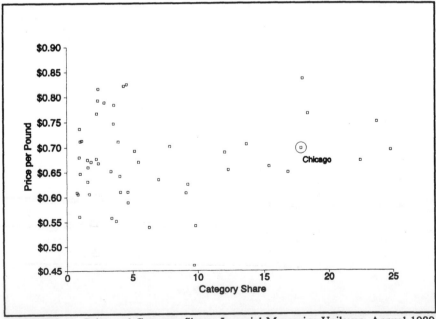

FIGURE 5.11 Price and Category Share, Imperial Margarine-Unilever: Annual 1989

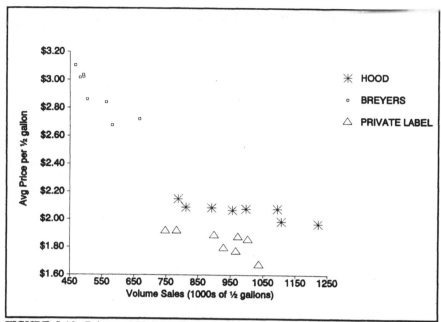

FIGURE 5.12 Price and Volume Sales, Ice Cream-Boston: Quarterly, 1988-1989

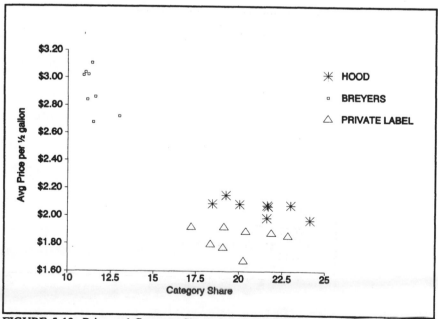

FIGURE 5.13 Price and Category Share, Ice Cream-Boston: Quarterly, 1988-1989

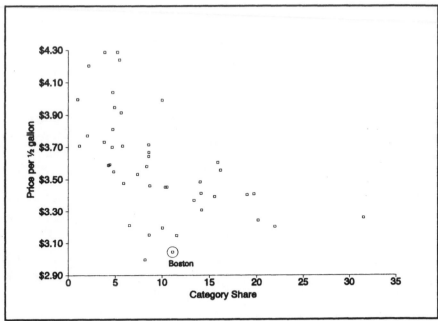

FIGURE 5.14 Price and Share, Breyer's Ice Cream-Philip Morris: Annual 1989 Data

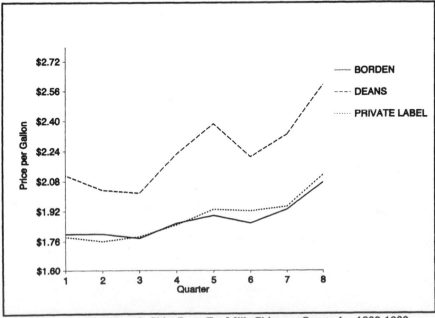

FIGURE 5.15 Price Trend, Skim/Low Fat Milk-Chicago: Quarterly, 1988-1989

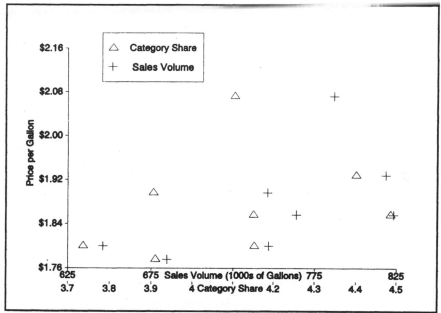

FIGURE 5.16 Price and Volume, Price and Share, Borden Skim/Low Fat Milk-Chicago:
Quarterly, 1988-1989

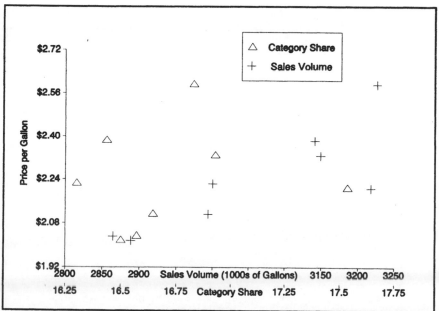

FIGURE 5.17 Price and Volume, Price and Share, Dean's Skim/Low Fat Milk-Chicago:
Quarterly, 1988-1989

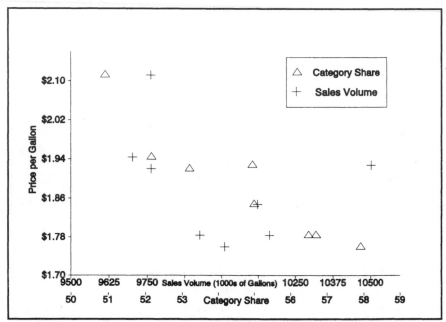

FIGURE 5.18 Price and Volume, Price and Share, Private Label Skim/Low Fat Milk-Chicago: Quarterly, 1988-1989

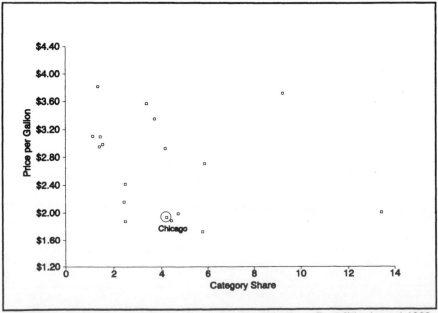

FIGURE 5.19 Price and Category Share, Borden Skim/Low Fat Milk: Annual 1989

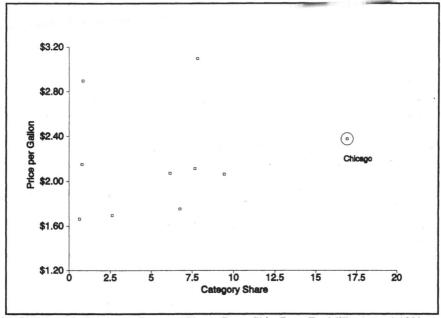

FIGURE 5.20 Price and Category Share, Dean Skim/Low Fat Milk: Annual 1989

Notes

1. This research was supported by Special Research Grant No. 91-34178-6330 with the Cooperative State Research Service, USDA, and by the Storrs Agricultural Experiment Station. This report is Scientific Contribution No. 1504, Storrs Agricultural Experiment Station, Storrs, CT 06269.

References

Baker, Jonathan B. and Timothy F. Bresnahan. 1988. Estimating the Residual Demand Curve Facing a Single Firm. *International Journal of Industrial Organization* 6: 283-300.

Connor, John M. and Everett B. Peterson. 1992. Market Structure Determinants of National Brand-Private Label Price Differences of Manufactured Food Products. *Journal of Industrial Economics* XL(2):157-171.

Cotterill, Ronald W. 1993. Introduction and Overview. In *Competitive Strategy Analysis in the Food System*, ed. Ronald W. Cotterill, 1-21. Boulder, Co.: Westview Press.

Cotterill, Ronald W. and Clement W. Iton. 1993. A PIMS Analysis of the Structure-Profit Relationships in Food Manufacturing. In *Competitive Strategy Analysis in the Food System*, ed. Ronald W. Cotterill, 23-43. Boulder, Co.: Westview Press.

Haller, Lawrence E. 1993. Branded Product Marketing Strategies in the Cottage Cheese
    Market: Cooperative versus Proprietary Firms. In *Competitive Strategy Analysis in
    the Food System*, ed. Ronald W. Cotterill, 155-177. Boulder, Co.: Westview Press.
Harris, Frederick. 1988. Testable Competing Hypotheses from Structure-Performance
    Theory: Efficient Structure versus Market Power. *Journal of Industrial Economics*
    36(3):267-280.
Parker, Russell C. and John M. Connor. 1979. Estimates of Consumer Loss Due to
    Monopoly in the U.S. Food-Manufacturing Industries. *American Journal of
    Agricultural Economics* 61(4):626-639.
Simmons, Tim. 1992. The European Model. *Supermarket News* 42(24):2.
Wann, Joyce and Richard J. Sexton. 1993. Analysis of Imperfect Competition in
    Multiproduct Food Industries: Theory and Application to California Pear Processing.
    In *Competitive Strategy Analysis in the Food System*, ed. Ronald W. Cotterill, 71-90.
    Boulder, CO: Westview Press.
Wills, Robert L. 1985. Evaluating Price Enhancement by Processing Cooperatives.
    American Journal of Agricultural Economics 67(2):183-92.

6

Vertical Quality Control Systems: A Potential Marketing Advantage for Cooperatives

Julie A. Caswell and Tanya Roberts

Many consumers are placing increased emphasis on the quality attributes of the foods they buy. These attributes may include freshness, taste, appearance, and convenience. More prominently, however, and carrying higher potential gains and losses to farmers and food processors, is the increased emphasis on the quality attributes associated with safety. Experts typically categorize safety concerns into six areas (in descending order of importance): microbial contamination, nutritional imbalances, naturally occurring toxicants, environmental contaminants, pesticide residues, and food additives. We group these concerns into two categories: food safety and nutrition. A food product's safety and nutrition profile is often affected by decisions made and actions taken at multiple stages in the chain of production, processing, and distribution. The focus here is on whether cooperatives may have an advantage over the investor-owned firms (IOFs) they compete with in designing and operating vertical quality control systems that aim to produce products with improved safety and nutritional attributes.

In recent experience, agricultural marketing cooperatives, as well as other food handlers and processors, have mostly seen the downside of consumer concerns about food safety and nutritional quality. Disruptions in demand for commodities such as milk, apples, chicken, and eggs after safety concerns became widespread have affected cooperatives and IOFs alike. Here we explore whether this downside risk can be turned into a marketing advantage and whether agricultural cooperatives may be in a special position to do so. Similarly, we analyze how quality assurance initiatives may be incorporated into long-term marketing plans rather than being the subject of short-term crisis management.

Delivery of food products with improved safety or nutrition profiles often requires extensive coordination between the producer, first handler, processor, and retailer levels. Capturing the value-added to consumers associated with such products requires effective communication of the relevant attributes through labeling and advertising. We explore whether establishment of vertical quality assurance systems that allow such marketing is a viable new strategic direction for cooperative management.

Food Safety's Downside Risks

A brief review of several food safety incidents will serve to illustrate the downside risks familiar to all food system players. Such incidents can be very costly and, as they occur, we are continuing to learn more about their dynamics. The cranberry "scare" of 1959 is a frequent starting point for discussing food safety incidents. The incident began shortly before Thanksgiving when the Secretary of Health, Education, and Welfare announced that cranberries might contain residues of the herbicide amino triazole (Brown 1969). Brown's analysis of the purchases of a panel of 300 families in Atlanta, Georgia showed that "per capita purchases of processed cranberries dropped 26 percent during 1959, compared with 1957-58, but regained the 1957-58 level during 1960-62 (p. 676)." He also speculates that increased advertising subsequent to the scare was effective in restoring the market.

The sources of costs associated with food safety incidents are numerous. These costs, presented as potential savings, are outlined in Table 6.1. They include acute and chronic costs to be avoided by farmers, processors, and retailers from improving food safety. Todd (1987) documents costs from three of these sources: product seizures, fines, and court settlements to injured parties for several incidents occurring during the 1960s and 1970s. In the 1980s, food safety incidents occurred on a regular basis. Many have lessons to teach about the nature of such incidents. Examples include:

• Heptachlor contamination of Hawaiian milk in 1982. Smith, van Ravenswaay, and Thompson (1988) estimate that Class I milk sales decreased by 29 percent in the months directly following the contamination and associated recalls. This decrease occurred despite the fact that all milk on sale was presumably safe. Smith *et al.* also found that while negative media coverage was important in explaining sales losses, positive media had no effect on consumer purchases. They conclude that:

> Although government officials repeatedly assured Oahu consumers that milk remaining on store shelves was safe, consumers appeared to have heavily discounted their statements. This implies that public statements by producers or government to assure the public may be of limited usefulness, at least in the short run (p. 519).

TABLE 6.1 Potential Sources of Industry Cost Savings from Improving Food Safety

Type of Costs	Farm	Processor	Retailer
Acute Costs Avoided			
Reduced demand for products	X	X	X
Product recall		X	X
Fines		X	X
Investigation costs		X	X
Cleanup costs and plant closings		X	X
Liability suits		X	X
Food handler illness	X	X	X
Chronic Costs Avoided			
Mortality of animals	X		
Morbidity of animals	X		
Reduced growth rate	X		
Reduced feed efficiency	X		
Worker illness	X	X	X
Food spoilage losses		X	X
More costly processing techniques		X	X

Source: Roberts (1992).

• Salmonella contamination of milk from a Jewel Company plant in the Chicago area in 1985. A faulty connection is believed to have inadvertently allowed mixing of raw and pasteurized milk in the supermarket company's private label milk processing plant. Over 16,000 culture-confirmed cases of salmonella infection occurred, with two surveys estimating that the number of people actually affected was from 169,000 to 198,000 (Ryan *et al.* 1987). While no economic analysis is available of the depth and duration of sales losses in this incident, press reports indicated that Jewel would ultimately pay $25-$50 million in compensatory damages to approximately 15,000 claimants (Natschke 1987).

• Alar residues in apples and processed apple products, 1984-1989. The Alar incident was particularly important in framing many food system participants' attitudes toward food safety issues. Although tolerances for Alar in food products were, after lengthy consideration, revoked by the Environmental Protection Agency in 1990, many people persist in believing that Alar was a safe chemical drummed off the market by overzealous environmental and consumer advocates. The manner in which the Alar controversy came to a head in 1989 (a Natural Resources Defense Council (NRDC) report, a *60 Minutes* segment, and other media events) was spectacular but careful research conducted thereafter by van Ravenswaay and Hoehn (1991a) for the New York City/Newark market indicates that:

Information about the risks of Alar reported in the press was found to have had a significant impact on demand for fresh apples ... this impact was found to have occurred much earlier than anticipated. Significant declines in sales were observed as early as the initial press reports [in 1984] of EPA's revised assessment of Alar's risks. In contrast, events that were initially anticipated to be responsible for significant declines in fresh apple sales (i.e., the NRDC report) were not found to be statistically significant when considered in isolation from other events (p. 170).

A clear lesson of this incident is that prompt attention to risk concerns as they arise will place an industry in a stronger position and tend to minimize long-term costs, while not addressing concerns may result in dramatic sales losses. Many, however, appear to be learning less productive lessons from this incident.

Timely attention to potential food safety problems may help avoid these types of food safety incidents. Vertical quality control systems can serve to produce products positioned to respond to consumer demands for food safety. They can also, of course, be viewed simultaneously as cost control mechanisms. While the future is difficult to predict, the large societal costs estimated to be associated with inadequate food safety indicate that the issue will gain in prominence in the future. For example, a partial estimate of the costs in the United States associated with foodborne disease is from $4.5-$12.4 billion a year (see Table 6.2). Rising standards of living and the fact the demand for food safety appears to be income elastic are other forces likely to enlarge the market for food safety in the future.

Marketing Opportunities

Vertical quality control systems serve offensive and defensive purposes (Caswell and Johnson 1991). The defensive purposes are better developed and understood. They focus on risk management and involve, for example, avoiding liability for selling hazardous products and drops in demand due to safety incidents. At the same time, offensive uses of quality control form a basis for product differentiation, which may either implicitly or explicitly tout the product's safety or nutritional attributes. Marketing based on safety and nutrition emphasizes an attribute approach to consumer demand. Altering the mix of attributes a product offers, particularly its safety features, will hopefully increase demand for the product. Such specific safety marketing is a relatively new phenomenon in the United States. Since food and drug regulations have been in place, the food marketing system has largely operated on the premise that all food available for sale was safe. There was no basis for differentiation because there was no real variation in safety levels across brands. This approach to safety, with governmental and private regulatory bodies setting a quality floor, which is also in effect a ceiling, is familiar and widely accepted.

TABLE 6.2 Annual Estimated Foodborne Disease Costs

Type of Cost	Annual Estimates (Billion $)
Consumer Costs	
Acute illnesses:	
Bacterial diseases	3.5-4.8
Parasitic diseases	.2-6.6
Viral diseases	?
Fungal toxins	?
Chronic illnesses:	
Bacterial diseases	?
Parasitic diseases	?
Viral diseases	?
Fungal toxins	?
Disutility of illness	?
Defensive expenditures	?
Industry Costs	
Outbreak costs	.4-.6?
Defensive costs	.1?
Public Health Sector Costs	
Federal	.3
State	?
Local	?
TOTAL	$4.5-12.4+

Source: Roberts (1992)

Over the past decade or so, however, markets have slowly begun to develop for products with safety profiles that exceed the government standard. This requires thinking of the government standard as only a floor and not a floor/ceiling. Some firms have used product formulation and advertising to differentiate their products based on food safety. To date, this process of attribute marketing is more advanced for nutritional characteristics than for food safety attributes. A visit to the local grocery store will reveal extensive nutrition-related marketing in numerous product categories as well as some safety-related marketing (Kaufman and Newton 1990).

In general, successful attribute marketing requires the market itself to have several features (see Table 6.3). On the demand side, consumers must be aware of and have effective demand for the attribute. On the supply side, firms must

TABLE 6.3 Characteristics of More Developed Attribute Markets

---

Consumer Awareness of Attribute's Importance
Effective Consumer Demand for the Attribute
Attribute-Based Product Formulation
Attribute-Based Product Differentiation
Relatively Well-Defined Regulatory Environment
    Standards
    Labeling
    Advertising

---

engage in attribute-based product formulation and be able effectively to communicate their differentiation messages to consumers through brand names, symbols, certifications, or other means. Doing this requires a relatively well defined and stable regulatory environment that sets clear standards, which serve as a basis of comparison, and regulates communication through labeling and advertising. The market for nutritional attributes currently has, or is near to having, all of these features. The market for other safety features is at an earlier stage of development. As we discuss below, this level of development makes more complex the strategic issues facing cooperatives and investor-owned firms that want to actively engage in this market.

There are some good reasons why the private market for food safety attributes has been slow to develop. Many people are uncomfortable with markets for food safety, thinking that the societal benefits of "every food a safe food" far outweigh the advantages of increased consumer choice. A specific benefit of this approach is its contribution to strong consumer confidence in the food supply. While some industry analysts think promotion of one brand for its safety attributes impugns the quality of other products in the category, this was not the case for new nutrition-based formulations of existing products such as low-fat mayonnaise or low calorie sodas. The original versions were still offered. Safer foods will likely be more expensive and targeted to high-risk or high-concern groups initially. The slow growth of the organic market may be a good model.

A further barrier to the development of private food safety markets is that they are characterized by asymmetric information, with the producer having better information than the consumer (Akerlof 1970). In this situation, unregulated markets may perform poorly in delivering food safety (Zellner 1988). Increased reliance on markets requires development of information mechanisms, such as public education and labeling, that decrease the asymmetry and improve the consumer's decision-making ability. Advances being made in this area make well-functioning food safety markets plausible. Barring much more active government regulatory action, the market for safety-differentiated products is likely to continue to evolve, sometimes slowly and occasionally in spurts. This evolution presents the marketing opportunity for cooperatives discussed here.

Is there a substantial market for products with improved safety or nutrition attributes? For marketers, two demand characteristics are key: the size of the potential market and consumers' willingness-to-pay for improved attributes. Research by McGuirk, Preston, and McCormick (1990) sheds some light on both issues (see, also, McCormick, McGuirk, and Preston 1989). Using Food Marketing Institute survey data, they found that about 39 percent of consumers are very concerned about food safety and have limited responsiveness to price-oriented promotion. Another 38 percent were highly concerned about safety and nutrition and were responsive to price-oriented promotions. A final 24 percent were not concerned and were not sensitive to promotional efforts. McGuirk *et al.*'s (1990) research suggests that the market for safety- and nutrition-oriented products is of very substantial size, with a little over half of the concerned consumers having lower levels of price sensitivity. This may translate into a substantial willingness-to-pay for products with improved attribute profiles or willingness-to-purchase them over other products. Demographic and attitudinal changes further suggest that the market for safety and nutrition-improved products will expand in the future. Finally, the market is likely to be further augmented by new scientific and medical knowledge that identifies individuals with higher risks associated with food safety concerns and recommends specific food purchasing and consuming habits for them.

Consumers' willingness-to-pay for improved products necessarily depends on their current stock of knowledge about the risks and benefits associated with competing products. Where consumers' information is imperfect, as it usually is to some degree, there may be no direct correspondence between willingness-to-pay and real risks and benefits. This wedge is a source of frustration to the food industry and has had unfortunate consequences in shaping attitudes toward food safety assurance. That consumers may hold some unsubstantiated beliefs about food safety has sometimes been grasped as a straw man to discount all food safety concerns. Equally unfortunate has been some firms' willingness to market products based on false consumer perceptions. The more difficult, and ultimately more rewarding long-run strategy, is to identify real risks, develop risk reduction/quality assurance programs and new products that address those risks, and communicate these to consumers. This strategy combines defensive and offensive elements and has the greatest probability of success in building consumer confidence and product/brand loyalty.

While consumer willingness-to-pay and willingness-to-purchase are the key issues for marketing decisions concerning safety and nutritional attributes, it is important to understand that government regulatory decisions have a separate basis in full societal benefit/cost analyses. Thus government regulations may be promulgated requiring private costs greater than consumer willingness-to-pay placing farmers and food processors in a squeeze. Additionally, changes in government regulations may from time to time undermine cooperatives and IOFs' efforts to establish safety-based product differentiation. For example, apples sold as Alar-free lost this basis of differentiation when Alar was removed from the market. Producers and processors who wish to market based on safety attributes must identify those concerns 1) with an adequate willingness-to-pay base to support marketing costs and 2) where future regulatory actions are not likely to erode the value of differentiation. Marketing to high-risk subpopulations may meet both criteria and avoid triggering a general public demand to raise the food safety floor for everyone. Niche marketing for safety also may be particularly profitable. In the general case of a new product introduction, the new product is attempting to appeal to people's demand for variety, an existing product category will be shared by this new product, and sales of established products will likely remain high. In the niche market for safety, more is asked of the consumer, namely to replace all or most of their product purchases with this new, "safer" product. Overall, vertical quality control systems are a substantial defensive and offensive opportunity for food growers and processors.

Vertical Coordination Issues

Vertical coordination between food production, processing, and distribution takes place through a variety of mechanisms ranging from the spot market to vertical integration, with various forms of contracting and cooperation in between. Spot markets generally do not work well in producing quality attributes, particularly levels of quality above a floor level. On the production side, farmers must often change their inputs, production practices, and/or handling and storage practices in order to produce products with improved safety or nutritional attributes. These changes frequently involve increased risk as the new practices are tried out and, frequently but not always, higher costs. Farmers need more assurance than a spot market can offer that their efforts will be remunerated. On the processing and marketing side, cooperatives and IOFs engaged in marketing products with improved profiles need reliable sources of supply at reasonably predictable prices. Since supply sources may be limited, a spot market is unlikely to meet these needs. At the other end of the spectrum, vertical integration internalizes production decisions and turns coordination decisions over to a management team. Quality control in some vertically integrated markets encompasses serious food safety concerns. A prime example is the performance of the broiler industry in controlling foodborne pathogens in its products. In fact, all food animal and seafood products can trace some microbial contaminants back to pre-harvest origins.

Our focus is the potential comparative advantages that marketing cooperatives may have in regard to safety and nutrition quality assurance. The principal comparison is to various forms of contracting employed by IOFs and, secondarily, to spot markets. Actions that affect food safety or nutrition may occur, as noted, at any stage of the input, production, processing, and distribution chain. Three major sources of potential comparative advantage for cooperatives are the ability to:

- Influence input and production practices to yield foodstuffs with improved safety or nutritional attributes.
- Organize food handling and processing practices to yield products with improved safety or nutritional attributes.
- Credibly communicate quality assurance programs to consumers.

Key issues are whether cooperatives have different incentives and capabilities to operate successful vertical quality control systems.

A comprehensive analysis of cooperative incentives regarding production of the safety and nutritional attributes of food is not available. Use of quality standards has probably received the most attention in connection with marketing orders and their effects on market operation and performance (e.g., Thompson and Lyon 1989, Carman and Pick 1990). These standards involve grades, sizes, and maturity (Armbruster and Jesse 1983), which only indirectly, if at all, relate to safety and nutrition. The following points are intended to suggest the lines of analysis likely to be fruitful in pursuing cooperative incentives and capabilities in this area:

- Common property, public good aspects of food safety (Caswell 1990b). Public confidence in the safety of the food supply is jointly produced (and used) by farmers, processors, and government regulators. This confidence is subject to free rider abuse by individuals who believe they can put low quality products on the market without suffering much individual harm. Cooperatives may be able to institute incentives that reduce or eliminate such free rider problems.
- Differential transaction costs (Staatz 1989). Most areas of food safety and nutrition require action on both production and processing levels. Cooperatives may have advantages due to lower transaction costs in negotiating vertical quality control relationships. This may be particularly true due to the large uncertainties and information gaps often associated with efforts to affect food safety and nutrition attributes. Much is likely to depend, however, on the structure of producer, first handler, and processor markets. It should also be noted that cooperatives may actually face higher transaction costs in negotiating changes in production practices with their members. They may also lose first-mover advantages because extra time is needed to negotiate changes.

Where change in farm-level practices is required to produce products with improved attributes, a major strategic question facing the cooperative is what type of incentive system to use to stimulate those changes. For example, should the incentive be in the form of a price premium or penalty? There may be instances where the cooperative wants to prescribe a higher standard on all production rather than handling various levels of quality. Strong cooperative leadership to explain why the change is needed would likely be required in this situation. The following section further explores the marketing process for food safety and nutrition and possible cooperative strategies in this area.

The Process of Marketing Food Safety and Nutrition

The marketing of food safety and nutritional attributes as present or incorporated into products is a three-step process. It involves identifying important quality attributes and their critical control points developing vertical linkages and processing capacities to produce products with desirable quality profiles at reasonable costs; and implementing strategies that effectively communicate quality assurance to the consumer.

Identifying Important Attributes and Critical Control Points

The key strategic decision in designing vertical quality control systems is which attribute or set of attributes to select. This task is often seen as particularly daunting in the arena of food safety and nutrition. However, many safety concerns are routine and on-going and already the focus of considerable quality actions by food producers and processors. For other issues, despite the perception that these types of issues explode on the scene, emerging concerns are very often widely recognized long before any public relations explosion occurs. An example is Alar where four years intervened between the initial concerns being raised and the explosion that accompanied a *60 Minutes* report and other media coverage. The issue of *Salmonella* in food has been before us since at least a National Academy of Sciences report in 1969. The expertise is at hand to identify a short list of major safety and nutritional concerns for every food commodity. For long-term strategic purposes, cooperatives and IOFs alike are well-advised to focus on real risks versus fad concerns. They are also well-advised, however, to not take a "we know best" attitude toward consumers' perceived risks when they differ from experts' lists of real risks. There may be legitimate reasons for differences in rankings that make consumers' current concerns a solid basis for product differentiation. In addition, identification of attributes to focus on must be based on a thorough understanding of the regulatory environment, particularly established and likely future standards.

A secondary identification issue is more difficult. That issue is: Which of the risks should be the focus of defensive and/or offensive activity? This decision must include consideration of the benefits and costs to the operation from pursuing changes. Benefits can be measured as costs avoided and marketing advantages gained. Costs are closely associated with hazard analysis and determination of the critical control points for a particular safety problem. To be a successful target, a cost-effective means of control must be within the cooperative's scope of vertical operation. Part of the decision process at this stage is determining what level of safety or nutrition is practical as a goal. A short list of safety targets might include presence of foodborne pathogens, pesticide residues, fat content, and, the subject of our case study, animal drug residues.

Production, Procurement, and Processing Strategies

If cooperatives have an advantage in vertical quality control, and in marketing based on it, the ability to coordinate farm level decision-making with the cooperative's procurement and processing strategies is likely to be one major source. In other words, a cooperative's transaction costs in this activity may be lower than those of spot markets or contracting used by IOFs. Cooperatives may have an advantage in:

- Identifying and communicating the common good to be gained by the group through changes in production and processing practices.
- Delivering member educational programs that introduce and promote changes in production and processing practices.
- Designing and enforcing standards (for feed or other inputs) required to meet final product goals.
- Designing incentive systems (e.g., price premiums or penalties) that encourage switching to new practices.
- Providing support mechanisms for producers in newly emerging markets, e.g., marketing cooperatives for organic produce and drug-free veal (Borst 1991).

For cooperatives and IOFs alike, design of production, procurement and processing strategies is complicated by the need to make these systems compatible with the already existing extensive food safety and nutritional regulations. In some cases, the cooperative's incentive system may actually be at odds with a governmental regulatory system that is not providing good safety or nutrition incentives. For example, van Ravenswaay and Bylenga (1991) found that the Food Safety and Inspection Service (FSIS) had a continuing problem with antibiotic and sulfa residues in bob veal calves in the 1970s. In 1984, the CAST (Calf Antibiotic and Sulfa Test) system was initiated, which used a more rapid test and a higher sampling rate. However, residue levels did not decline and producers had an economic incentive to falsely certify their calves as residue-free. This occurred because the sampling rate was only 3 percent for certified versus 77 percent for non-certified calves. In addition,

TABLE 6.4   Economic Incentives in FSIS Regulatory Programs Identified by van Ravenswaay and Bylenga (1991)

Incentives
Condemn Violative Carcasses
Increase Testing Rate to Detect a Higher Percentage of Violators
Require Follow-Up Testing for Violative Producers Before Future Deliveries to the Plant
Imposing Penalties for Being Out-of-Compliance, Including Penalties for Marketing Unidentified Calves

slaughter plants did not charge producers for the loss of condemned calves so producers faced no penalty. In fact, production costs were $10 lower per bob veal calf if they did not comply with the law.

While FSIS was taking action to reduce violations by (1) condemning violative calves and (2) increasing the probability of detecting violators by increasing the testing rate, van Ravenswaay and Bylenga recommended further economic penalties of (3) requiring follow-up testing for violative producers before future calves could be slaughtered and (4) imposing penalties for being out of compliance. This example is useful in highlighting the types of economic incentives used by FSIS in its regulatory programs (see Table 6.4). Interactions between these systems and private cooperative or IOF systems will be a key to the latter's success.

Design of incentive systems is complex. Coaldrake and Sonka's (1991) case study of efforts to design incentive systems to encourage the production of lean pork shows a series of economic considerations and tradeoffs which included (1) the capital cost of equipment to detect leanness; (2) the speed of the monitoring machine, which could have secondary impacts on line speed; (3) the cost of administering the program to pay for each hog according to its leanness; (4) the effects on other carcass attributes such as the probability of Pale, Soft, Exudative (PSE) pork; and (5) the on-farm tradeoffs such as a possible effect of leanness on animal health. Taking all five potential factors into account when designing a new carcass merit system is complex. Some companies are experimenting with contractual production systems, others are tying their system to importation of breeding stock or other genetic improvements. Morrell is currently using the Fat-O-Meter in its carcass merit program. The Indiana Packing Company has installed TOBEC and plans to create a lean meat payment schedule (Coaldrake and Sonka 1991). This case study is suggestive of the types of challenges cooperatives are likely to face in designing incentive systems.

A cooperative disadvantage may lie in its member ownership organization. Changes in production practices that are resisted by farmers or growers may be difficult for the cooperative to promote through changes in product standards or implementation of price differentials. In this case, cooperatives may lag rather than lead the market.

Communicating Quality Assurance

Where vertical quality control programs are put in place for defensive purposes, communication regarding them will normally be limited to real or potential crisis situations. This function is usually the purview of a crisis management team, with the program's purpose being to prevent crisis situations from arising and to respond quickly and reassuringly should they occur. The communication function differs where vertical quality control is undertaken as a basis of product differentiation. Here its goal is to generate sales and capture the value-added to consumers from the improved product attributes. The major communication vehicles are advertising and product labels.

Strategy is likely to differ markedly depending on whether the cooperative actively sells in the final consumer market or supplies inputs to IOFs. In the former case, the cooperative directly controls the communication with the consumer and enjoys the payoffs from marketing products with improved attributes. In the latter case, the cooperative must negotiate "identity retention" for its inputs with the IOF, which has direct control over communication with consumers.

The use of advertising and labeling to convey product attributes is closely regulated in the United States. In addition, a multitude of grades and standards exist that regulate product formulation and presentation. Attribute differentiation is based on viewing this regulatory framework as setting rules of communication and a *floor* under product quality. For most nutritional attributes, products may be marketed with higher quality levels provided their advertising and labeling is not deceptive and makes all required disclosures. Under new legislation, nutritional labeling is becoming much more closely regulated for products under both Food and Drug Administration (FDA) and United States Department of Agriculture (USDA) jurisdiction. These new labeling requirements, which take effect in 1994, will have significant impacts on product formulation and advertising (Caswell and Padberg 1992).

Parallel developments are occurring in regard to labeling safety attributes, although this area is still in its infancy. The greatest activity has occurred regarding labeling of production and processing technologies, e.g., development of standards for organic products. Another example is the USDA's final rule on poultry irradiation, which specifies acceptable label messages (9 CFR Part 381, 21 September 1992, p. 43588). Several states are considering whether and how to label milk from cows treated with supplemental BST and analyzing such labeling's likely effects should the treatment be approved for widespread use in the dairy industry (Preston, McGuirk, and Jones 1991; Caswell 1990a). We expect labeling of production practices to develop rapidly over the next decade.

TABLE 6.5 Cooperatives' Share of Sales by Dairy Product Category, 1987

Product Category	% Cooperative Sales
ALL FARM MILK SALES	76%
PRODUCT SALES	
Dry Milk	91%
Butter	83%
Cheese	45%
Packaged Fluid Milk	14%
Cottage Cheese	13%
Ice Cream/Ice Milk	8%

Source: Ling and Roof (1989).

Cooperatives may have an advantage in vertical quality control if they have more credibility than IOFs in communicating quality claims to consumers. A source of this credibility might be farmers and cooperatives' positive image with the public, perhaps based on the land stewardship role farmers fulfill. The strength of this credibility will be in comparison to that of IOFs, private contract certification programs, and government certification systems. Work has recently been initiated on which organizations are viewed as credible sources of safety information (see, e.g. van Ravenswaay and Hoehn 1991b). Cooperatives may be able to develop an advantage in using existing labeling devices (e.g., government approved label claims) or in creating cooperative quality seals that gain acceptance through advertising and labeling. For marketers, of course, the bottom line focuses on the costs of differentiation, including changing product formulation and communicating to consumers, compared to consumer willingness-to-pay for the quality attributes being built into the product or gains in market share.

A Case Study: Quality Control in the Dairy Industry and Animal Drug Residues in Milk

A variety of food safety and nutrition issues are viable foci for vertical quality control systems operated by cooperatives. As a case study, we explore the experience and possible future developments regarding quality control in the dairy industry, particularly the issue of drug residues in milk. Drug residues are an especially interesting example because they have not yet attained widespread consumer awareness but are the focus of new regulatory efforts. This case study is also particularly relevant because of the large presence of cooperatives in the dairy industry. In the United States, cooperatives market 76 percent of all milk, 91 percent of dry milk, 83 percent of butter, and 14 percent of packaged fluid milk (see Table 6.5). Thus cooperatives have a large stake in the dairy industry and the industry, in turn, has a large stake in quality issues such as animal drug residues.

Demand for and Supply of Milk Quality Control

Quality control is of paramount importance in the milk industry. Product differentiation based on quality and safety has a long history in the dairy business. At the turn of the century every state had a "sanitary" dairy. "Purity" was also often used and traces of both names linger in the Northeast. Companies pasteurizing their milk used these names to signal reduced bacteria counts, which meant safer, longer lasting milk. Today milk is a more uniform product and flavor and sanitation differences are small. In evaluating milk quality, *Hoard's Dairyman* (1992) recently stated that:

> By far the most important measure of milk quality, flavor is what consumers use to judge milk and dairy products. Milk having good flavor is pleasing and slightly sweet to the taste and has good "mouth feel." Milk having good flavor does not leave a distinct aftertaste. Naturally, milk with good flavor has no off-flavor (p. 2).

Hoard's Dairyman's complete list of important attributes on which milk quality should be evaluated includes:

- Flavor.
- Odor.
- Color.
- Acid Degree Value.
- Sediment.
- Temperature.
- Drug Residues.
- Somatic Cell Count.
- Bacteria Count.
- Preliminary Incubation Count.
- Added Water.

Quality tests currently exist to one degree or another for each of these attributes.

Demand for high quality milk and dairy products, including low-fat and nonfat products, is likely to remain strong in the future. Flavor, quality, and healthfulness are very likely superior goods, with an income elasticity greater than one so that demand for these attributes will increase over time with rising incomes. Another factor likely to spur demand for products with specialized safety and nutritional attributes is new scientific knowledge that allows us to identify high-risk consumers. For the dairy industry, these consumers are pregnant women, cancer and AIDS patients, persons with organ transplants, and persons on corticosteroid therapy, all of whom are at increased risk of listeriosis because of their vulnerable immune systems (Schuchat *et al.* 1992, Pinner *et al.* 1992). Combined with rapid tests that will be available in the future for *Listeria* and *Yersinia*, both an identified market segment and a way to serve them will be available. An example of a product category where these changes are likely to have significant effects is frozen desserts, where an estimated 3% of the product has a *Listeria* contamination problem. Current tests for *Listeria* take 7-8 days making control strategies difficult to monitor.

A second example of an expanding niche market is acidophilus milk for the increasing number of persons who realize they are lactose intolerant. This market is also increasing because population groups with higher rates of lactose intolerance, such as older European Americans, African Americans, and Asian Americans, are growing faster than the general population.

Cooperatives and IOFs currently employ a range of economic incentives to encourage delivery of safe, high-quality milk. As shown in Table 6.6, these incentives focus primarily on somatic cell, standard plate, and preliminary incubation counts. They include premiums, penalties, or both ranging in size but generally representing a percentage incentive of less than 5 percent of price. These type of incentives can lead the way for incentives based on other safety and quality attributes. Such incentives will need to be designed in tandem with governmental testing requirements. The ability to put new incentives in place will also depend on the availability and cost of rapid tests for the quality attributes being monitored.

Background on the Animal Drug Issue

The Food and Drug Administration (FDA) sets residue tolerances for animal drugs in foods. These tolerances range from 0 to 10 parts per million (ppm) and use the most sensitive effect in the most sensitive species as a predictor of the threshold for any effect of the residue in humans. FDA's regulations include approval of animal drug uses shown to be effective against specific diseases in specific animals, maximum drug doses allowed, and withdrawal periods before marketing product from treated animals. Several concerns regarding animal drug use are before the industry, regulators, and the public:

• General extra-label (i.e., illegal) use of animal drugs in the United States (General Accounting Office 1992). This use includes doses above approved levels to combat resistent bacterial pathogens causing animal disease, non-approved disease/animal combinations, and use of drugs not approved for any food animal use. Legislation is pending in the U.S. Congress regarding such drug use.

• Anticipated European Community (EC) regulation of drug residues in food that sets standards stricter than those of the United States. The Community will set Maximum Residue Levels (MRLs) for all animal drugs that are permitted in food. Differences in standards are likely to cause trade conflicts. They will also likely focus attention on FDA's approach in setting tolerances (similar to the EC's MRLs), which relies on determining the No Effect Level (NOEL) for

TABLE 6.6 Current Economic Incentives for Milk Safety and Quality

Cooperative/Location	Incentive/Penalty ($ per Hundredweight)	Criteria (Cell Counts per Milliliter)
Wisconsin & Minnesota (Co-op and private)	Up to $.74/cwt Premium	SCC<50,000, SPC 10-100,000
Calif. Milk Prod. Assoc.	Penalties Up to $2/cwt	SPC>25,000, Coliform>500
Darigold Farms	$.20/cwt Premium Penalty Up to $.50/cwt	SCC<100,000, SPC<10,000, PI<50,000
Western Dairy Co-op, Inc.	$.10/cwt Premium	SCC, SPC, PI
AMPI-Southern	$.05-.20/cwt Premium $.05-.20/cwt Penalty	SCC<400,000 SCC>600,000
Michigan Milk Prod., Inc.	Up to $.30/cwt Premium	SCC<100,000, SPC<10,000, PI<20,000
Milk Marketing, Inc.	$.25-1.00/cwt Premium	For Top Premium SCC<150,000, SPC<10,000, PI<10,000
Eastern Milk Producers	Premium Up to $.35/cwt Penalty Up to $.20/cwt	SCC<150,000, SPC<10,000 SCC>1,000,000, SPC>100,000
Agri-Mark	$.10/cwt Premium	SCC<300,000, SPC<10,000, PI<50,000
Dairymen, Inc.	$.10-.20/cwt Premium	

SCC = Somatic Cell Count.
SPC = Standard Plate Count.
PI = Preliminary Incubation Count.
Source: Data from Hoard's Dairyman (1992), p. 8.

humans. Two key scientific questions FDA asks are whether antibiotic residues in food increase the resistance of bacteria in the human gut and whether antibiotic residues in food decrease the barrier to human infections by changing the human gut flora.

• Biotechnology-based products have to be specifically approved by the EC whereas in the United States foods produced using biotechnology will follow traditional FDA regulatory tracks (Gladwell 1992).
• Nitrofurans were recently banned from animal feeds by FDA. Penicillin, oxytetracycline, and chlortetracycline are on FDA's back burner.
• Development of organic standards as required by the 1990 Farm Bill. These standards could permit marketing of organic milk or meat, although how the standards would address drugs used to treat sick animals has to be determined. An additional question is whether feeds will have to be produced without herbicides or pesticides for 3 years before being fed to animals in organic production.

Animal Drug Residues in Milk

Concerns about animal drug residues in food products have been circulating for some time (see, for example, General Accounting Office 1987). Recent concern regarding milk was stimulated by surveys in late 1989 by the *Wall Street Journal* and the Center for Science in the Public Interest (Ingersoll 1989) finding several types of drug residues in a substantial number of samples tested. A General Accounting Office (GAO 1990) study subsequently concluded that FDA's survey techniques for animal drug residues were not adequate to demonstrate the safety of the milk supply.

In February, 1991, FDA initiated a new National Drug Residue Milk Monitoring Program for sulfanomides. This program was expanded in fiscal year 1992 and now includes tests for beta-lactams, novobiocin, chloramphenicol, and tetracyclines. Under the system, 15-20 milk samples are collected each week (500 samples/year), with samples split between Grade A and non-Grade A milk. More importantly, as of July 1, 1992, Appendix N of the Pasteurized Milk Ordinance requires that every bulk milk pick-up tank load be tested for beta-lactams (penicillin and penicillin-like drugs) prior to processing. Penalties for producers shipping antibiotic positive milk are substantial:

1st violation: Grade A permit suspended 2 days; mandatory completion of 10-point Milk & Dairy Beef Quality Assurance Program.
2nd violation: Grade A permit suspended 4 days.
3rd violation within 12 months: Grade A permit suspended for 4 days and action begun to revoke Grade A permit permanently.

However, a new GAO report on illegal drug residues in milk published in

August 1992 concludes that expanded state and federal testing and other regulatory efforts are inadequate (GAO 1992). The next Pasteurized Milk Ordinance Conference will take place in May, 1993 and the FDA could raise additional questions and suggest or require new tests.

Cooperatives' Stake in the Milk/Animal Drug Residue Issue

Cooperatives have a large stake in the issue of animal drug residues in milk because of their large share of the milk market and their "investment" in milk's public image as a wholesome, safe product. Many cooperatives are now responding to the new testing requirements by instituting clear penalties for so-called hot loads (loads contaminated with drug residues). *Hoard's Dairyman* (1992) quotes the policy of Milk Marketing, Inc., Strongsville, Ohio that:

> As of March 1, 1992, producers who contaminate a load of milk with drug residues are financially responsible to compensate others on that load for the loss of their milk and the cost of disposing of the milk (p. 14).

Michigan Milk Producers Association allows that "if members suspect a drug positive bulk tank, have it checked, and the tank load is dumped, they can receive 75 percent of the value of the milk for two such occurrences in a 12-month period (*Hoard's Dairyman* 1992, p. 14)." Members who contaminate a pick-up tank, however, face a schedule of penalties.

This animal drug residue example presents a major challenge to the premise of this paper that vertical quality control to assure safety and quality may be a viable new strategic direction for cooperatives. That challenge is: Why did the milk industry, and cooperatives in particular, not respond more aggressively to the animal drug residue issue *before* required to by new Pasteurized Milk Ordinance regulations? Potential means of responding could have been through testing programs and incentive systems to reward members who participate in the National Milk Producers Federation's 10-point Milk and Dairy Beef Quality Assurance Program for drug control. Properly designed, such a response could have been the basis of defensive gains (preparation for a public relations flurry, should it come) and offensive advantages in marketing milk with superior quality attributes.

Possible Directions for Marketing Dairy Products with Improved Safety and Nutrition Attributes

There are several potential marketing opportunities for milk with improved safety and nutritional attributes. These opportunities with regard to nutritional attributes such as fat content are well understood throughout the industry. Also well understood is the importance to consumer acceptance of product presentation. The product's name can be of particular importance in determin-

TABLE 6.7 Existing and Hypothetical Milk Labels Signaling Safety to Consumers

Label Examples	Comments
"Quality chekd"	Red check mark in circle, expand food safety tests, and advertise as above Federal standards.
Ultra-pasteurized to extend shelf-life & control foodborne pathogens	Comparable to label permitted on irradiated chicken.
Organic	1990 Farm Bill permits organic labeling, how will it be defined? Permit drugs to treat sick animals?
Meets highest Pennsylvania quality standards for safety, flavor, and shelf-life	Bakery products have PA label; PA checks milk quality at consumer purchase level.
+ Milk (safety+, flavor+, shelf-life+, and digestible+)	Augment Dairy Science Association 0-10 scorecard; (10 is highest) by increasing food safety tests.
Safety and quality money-back guarantee if indicator turns red (or if this is not the best milk you have ever tasted)	Potential use for time-temperature indicators.
Meets *Listeria* testing standards above FDA requirements	Niche product for pregnant women, cancer and AIDS patients, those with organ transplants or steroid therapy.
Cholesterol "modified" milk	Nutrition Labeling and Education Act has strict standards for "reduced" cholesterol, milk not likely to meet them.

ing product sales—ice milk and "sour" half-and-half did not sell well, but low-fat ice cream and low-fat sour cream are starting to sell very well. Acidified cottage cheese does not sell well, but "directly set" cottage cheese does. Table 6.7 presents a set of existing and hypothetical milk labels that could be used to signal quality attributes to consumers. They cover a range of quality attributes and combinations of quality attributes:

● "Quality chekd." This marketing association's label is already being sold in some Northeastern markets on a variety of dairy products. Products with this label have passed increased food safety tests and may advertise that their tests are above FDA safety standards.

- Ultra-pasteurized to extend shelf life and control foodborne pathogens. This claim would parallel the language FSIS permits on irradiated chicken, "irradiated to control foodborne pathogens (9 CFR Part 381, 21 September 1992, p. 43595)." The claims inform consumers of the food safety benefits associated with these two processes. FDA would have to be petitioned to allow such a label on ultra-pasteurized dairy products.

- Organic. Depending on how the organic standards are implemented, we could see provisions for organically produced dairy products and meats. Some of the issues include use of drugs to treat sick animals and whether any herbicides or insecticides can be used in the production of animal feeds.

- Meets highest Pennsylvania quality standards for safety, flavor, and shelf-life. Currently bakery products marketed in the Northeast often note Pennsylvania registration. Pennsylvania is the only state which monitors dairy products in consumer packages. This performance information by brand name and product is not available to the public, but it could be; and dairy products could be advertised as meeting Pennsylvania standards.

- + Milk. This is a hypothetical brand which also includes the acidophilus market for people who are lactose-intolerant. Skim acidophilus has great "mouth-feel" although it is somewhat sweet. Another flavor rating plan that could be used is the Dairy Science Association's 1-10 scorecard. Perhaps a new rating scorecard could be developed combining all four attributes: safety, flavor, shelf-life, and digestibility.

- Safety and quality money-back guarantee if indicator turns red (or it this is not the best milk you have ever tasted). The development of time-temperature indicators makes available a new technology to assure consumers that milk has been kept appropriately refrigerated in the marketing chain. The indicators turn color when temperature abused or the shelf-life is overextended. Since they cost $0.02-$0.05 each, gallons of milk may be a more likely use than quarts, although the smaller volume in quarts makes them more vulnerable to temperature abuse. Introduction of such time-temperature indicators would encourage better milk handling and lower refrigeration temperatures (38° F is optimum for quality and shelf-life but most supermarket dairy cases are now at 45° F). However, this would come at the price of higher energy costs. Some high-risk consumers would clearly benefit from introduction of time-temperature indicators since their exposure to foodborne diseases could thus be lowered.

- Meets *Listeria* testing standards above FDA requirements. Such a label would permit those at increased risk of listeriosis—pregnant women, cancer or AIDS patients, and those with organ transplants or steroid therapy—to consume cheeses, milk, and frozen desserts with more confidence.

- Cholesterol "modified" milk. This last label is targeted to those at risk of heart disease. Here the standards of the Nutrition Labeling and Education Act of 1990 are crucial. It is unlikely that milk could meet the strict standards for "reduced" cholesterol, but perhaps it could meet some lesser standard.

These labels are presented to stimulate thinking about how product attributes and labels could be modified to target products to specific consumers and create niche markets satisfying their demands for safety attributes. Some of the these attributes might appeal to the mass market as well. The feasibility of marketing on these bases is a ripe issue for study by cooperative strategic planners.

Barriers to Successful Operation of Vertical Quality Control Systems by Cooperatives

Several potential impediments exist to cooperatives' successful operation of vertical quality control systems and marketing of safety and nutrition attributes. A list of these includes:

- Pursuit of these systems may impugn the safety and quality of existing production practices and products.
- Difficulty in forecasting demand for attributes.
- Cost/payoff comparison may show market will not support these efforts.
- Design of production incentive systems may cause conflict.
- Retaining identity through the marketing channel may be difficult, particularly for cooperatives not involved in processing and further value-added activities.
- Difficulties with credible and cost-effective communication to consumers.

Future Directions

Demand for the safety and nutritional attributes of foods will increase in the future as incomes rise, vulnerable populations are further identified, and tests for these attributes become faster and less expensive. The increased demand will augment the market shares of products with improved safety profiles and may also garner them a price premium. This demand presents a marketing opportunity to cooperatives as well as to investor-owned firms. Cooperatives may have advantages over IOFs in some product markets because of: (1) possible lower transaction costs in negotiating farm-level incentives for providing specific safety attributes desired by consumers and (2) because they may be able to skirt free rider problems associated with lower product quality. However, to capture the value-added to consumers from these attributes, cooperatives will likely need to market more branded products. While we have proposed possible sources of cooperative advantage, several possible barriers to cooperatives serving these markets also exist. Where the balance lies will be determined by cooperatives' strategic marketing decisions over the next few years.

Notes

The views expressed are those of the authors and do not necessarily represent those of the United States Department of Agriculture. The authors appreciate information on dairy industry programs provided by Richard F. Fallert and Don P. Blagney, Economic Research Service, USDA and comments from Sylvia Lane and from Phil Kaufman and Betsey Frazao, Economic Research Service, USDA. An earlier draft of this paper was presented at the Northeast Regional Research Committee NE-165 Workshop *New Strategic Directions for Agricultural Marketing Cooperatives*, Boston, Massachusetts, June 24, 1992.

References

Akerlof, George A. 1970. The Market for "Lemons": Qualitative Uncertainty and the Market Mechanism. *Quarterly Journal of Economics* 84: 488-500.

Armbruster, Walter J. and Edward V. Jesse. 1983. Fruit and Vegetable Marketing Orders. In *Federal Marketing Programs in Agriculture: Issues and Options*, Walter J. Armbruster, ed., pp. 121-158. Danville, IL: The Interstate Printers and Publishers, Inc.

Borst, Alan D. 1991. *The State of U.S. Organic Producer Marketing Cooperatives in 1991*. Washington, D.C.: USDA, ACS, Staff Paper 91-S7, August.

Brown, Joseph D. 1969. Effect of a Health Hazard "Scare" on Consumer Demand. *American Journal of Agricultural Economics* 51(3):676-678.

Carman, Hoy F. and Daniel H. Pick. 1990. Orderly Marketing for Lemons: Who Benefits? *American Journal of Agricultural Economics* 72(2):346-357.

Caswell, Julie A. 1990a. Economic Impacts of Food Safety and Nutrition Marketing. In *Agricultural Biotechnology, Food Safety, and Nutritional Quality for the Consumer*, NABC Report 2, J. F. MacDonald, ed., pp. 174-180. Ithaca, New York: National Agricultural Biotechnology Council.

___. 1990b. Food Safety Policy Fights: A U.S. Perspective. *Northeastern Journal of Agricultural and Resource Economics* 19(2):59-66.

Caswell, Julie A. and Gary V. Johnson. 1991. Firm Strategic Response to Food Safety and Nutrition Regulation. In *Economics of Food Safety*, Julie A. Caswell, ed., pp. 273-297. New York: Elsevier Science Publishing Co., Inc.

Caswell, Julie A. and Daniel I. Padberg. 1992. Toward a More Comprehensive Theory of Food Labels. *American Journal of Agricultural Economics* 74: 460-468.

Coaldrake, Karen and Steven Sonka. 1991. NE-165 Case Study. Leaner Pork: Can New Sector Linkages be Formed? NE-165 Working Paper No. 26. Department of Agricultural Economics, University of Illinois, October.

General Accounting Office. 1992. *Food Safety and Quality: FDA Strategy Needed to Address Animal Drug Residues in Milk*. GAO/RCED-92-209, Washington, D.C., August.

___. 1990. *Food Safety and Quality: FDA Surveys Not Adequate to Demonstrate Safety of Milk Supply*. GAO/RCED-91-26, Washington, D.C., November.

___. 1987. *Imported Meat and Livestock: Chemical Residue Detection and the Issue of Labeling*. GAO/RCED-87-142, Washington, D.C., September.

Gladwell, Malcolm. 1992. Biotech Products Won't Require Special Rules, FDA Decides. *Washington Post*, 26 May, p. A4.

Hoard's Dairyman. 1992. Drug and Milk Quality Supplement. 25 May, pp. 2-15.

Ingersoll, Bruce. 1989. Dairy Dilemma: Milk is Found Tainted with a Range of Drugs Farmers Give Cattle. *Wall Street Journal*, 29 December, p. 1.

Kaufman, Phil and Dorris J. Newton. 1990. Retailers Explore Food Safety and Quality Assurance Options. *National Food Review* (October-November):11-15.

Ling, K. Charles and James B. Roof. 1989. *Marketing Operations of Dairy Cooperatives*. Washington, D.C.: USDA, ACS Research Report 88.

McCormick, A., A. M. McGuirk, and W. P. Preston. 1989. Marketing Food Safety: Toward the Development of Product Differentiation Strategies. *Southern Journal of Agricultural Economics* 21: 185.

McGuirk, A. M., W. P. Preston, and A. McCormick. 1990. Toward the Development of Marketing Strategies for Food Safety Attributes. *Agribusiness* 6(4):297-308.

National Academy of Sciences, National Research Council, Committee on Salmonella. 1969. *An Evaluation of the Salmonella Problem*. Washington, D.C.: National Academy Press.

Natschke, Patricia. 1987. Jewel May Pay Up to $50M to Salmonella Claimants. *Supermarket News*, 30 November 1987.

Pinner, Robert W. *et al.* 1992. Role of Foods in Sporadic Listeriosis. II. Microbiologic and Epidemiologic Investigation. *JAMA: Journal of the American Medical Association* 267 (15, April 15):2046-2050.

Preston, W., A. McGuirk, and G. Jones. 1991. Consumer Reaction to the Introduction of Bovine Somatotropin. In *Economics of Food Safety*, J. A. Caswell, ed., pp. 189-210. New York: Elsevier Science Publishing Company, Inc.

Roberts, Tanya. 1992. Estimated Foodborne Disease Risks and Costs in the United States. Paper presented at the Allied Social Science Associations meetings, New Orleans, January.

Ryan, Caroline A. *et al.* 1987. Massive Outbreak of Antimicrobial-Resistant Salmonellosis Traced to Pasteurized Milk. *JAMA: Journal of the American Medical Association* 258(22):3269-3274.

Schuchat, Anne *et al.* 1992. Role of Foods in Sporadic Listeriosis. I. Case-Control Study of Dietary Risk Factors. *JAMA: Journal of the American Medical Association* 267 (15, April 15):2041-2045.

Smith, M. E., E. O. van Ravenswaay, and S. R. Thompson. 1988. Sales Loss Determination in Food Contamination Incidents: An Application to Milk Bans in Hawaii. *American Journal of Agricultural Economics* 70:513-520.

Staatz, John M. 1989. *Farmer Cooperative Theory: Recent Developments*. Washington, D.C.: USDA, ACS, ACS Research Report No. 84, June.

Thompson, Gary D. and Charles C. Lyon. 1989. Marketing Order Impacts on Farm-Retail Price Spreads: The Suspension of Prorates on California-Arizona Navel Oranges. *American Journal of Agricultural Economics* 71 (3):647-660.

Todd, Ewen C. D. 1987. Legal Liability and its Economic Impact on the Food Industry. *Journal of Food Protection* 50(12):1048-1057.

van Ravenswaay, Eileen O. and Sharon Bylenga. 1991. Enforcing Food Safety Standards: A Case Study of Antibiotic and Sulfa Drug Residues in Veal. *Journal of Agribusiness* 9(1):39-53.

van Ravenswaay, E. and J. P. Hoehn. 1991a. The Impact of Health Risk on Food Demand: A Case Study of Alar and Apples. In *Economics of Food Safety*, J. A. Caswell, ed., pp. 155-174. New York: Elsevier Science Publishing Company, Inc.

van Ravenswaay, E. O. and J. P. Hoehn. 1991b. Consumer Willingness to Pay for Reducing Pesticide Residues in Food: Results of a Nationwide Survey. Staff Paper No. 91-18, Department of Agricultural Economics, Michigan State University.

Zellner, James A. 1988. Market Responses to Public Policies Affecting the Quality and Safety of Foods and Diets. In *Consumer Demands in the Marketplace: Public Policies Related to Food Safety, Quality, and Human Health*, K. L. Clancy, ed., pp. 57-73. Washington, D.C.: Resources for the Future, National Center for Food and Agricultural Policy.

## PART THREE

Strategic Alliances: Cooperative Marketing Agencies in Common

7

Cooperatives and Marketing Agencies in Common

Bruce J. Reynolds

Common marketing agencies (CMAs) have long been a useful form of organization for cooperatives. The Capper-Volstead Act specifically extends to them the same limited anti-trust protection as provided for direct farmer membership cooperatives. Part of their attraction also seems to be a combination of attributes that usually do not go together—locally controlled organizations with a system for market power and economies of size. There have been many successful CMAs and they will continue to be formed.

When cooperatives establish CMAs they are usually looking to achieve one or more of the following objectives:

1. Market power for negotiating price enhancements;
2. Economies of size in operations;
3. Information sharing, which includes price coordination; and
4. Market development, or value-added.

These are some of the factors that would be examined in the technical assistance work of the Agricultural Cooperative Service (ACS), but such studies would be remiss if they did not also point out that a merger alternative would attain these objectives more fully than a CMA would. However, a merger alternative provides a benchmark only in situations where cooperatives are contemplating coordination of their total sales operations. It is clearly not a benchmark for evaluating typical specialization of CMAs for ancillary services like by-products or institutional sales.

There are many weaknesses of CMAs, with probably the most widely recognized one being an inducement for free-rider behavior when trying to accomplish either market power or price coordination. Another is that local control of member associations does not always filter through to a CMA. In addition, CMAs are often valued as a transitional form, providing valuable experience that will eventually lead to merger, but there is a potential negative side to such a strategy. It is possible that some CMAs have become a way to avoid mergers, with the result that structural adjustments are delayed until it becomes too late to gain a competitive foothold in the market. Cooperative grain marketing has examples of CMAs for exporting that, in retrospect, may have functioned as delaying tactics.

The purpose of this paper is to examine the conditions that are appropriate and conducive to the effective workings of CMAs, and to discuss potential limitations to pursuing objective (4), market development. The term "market development" is used in this context to encompass such attributes as marketing innovations and targeted demand enhancement for member products. Enhancing the value or expanding the markets for member products is one of a cooperative's key objectives. It is a distinguishing feature from investor owned firms that procure agricultural products but have, if anything, a disincentive to foster demand for them, at least not before they have imparted some type of time, place or form utility to these raw commodities. "Value-added" is a useful term when measuring the contribution of these different utilities, but in terms of the operational goals of a cooperative, the notion of market development is preferable. Members and managers should not conceive of the cooperative as the primary point for value creation, but rather as an instrument or agent for augmenting the value of the members' product.

Interest in using CMAs for the market development objective is currently very strong, particularly among dairy cooperatives. However, examples of such application are slight. There are many examples of CMAs for the other three objectives, although the reasons cited by participants for success or failure are difficult to validate. Economic theory of cooperatives can help overcome this difficulty, as well as understand the constraints to coordination of market development. Economic reasoning should also be applied to identifying relevant differences between a producer cooperative and a CMA or any federated cooperative, other than the obvious fact that one has people and the other has organizations as direct members.

The potential of adapting recent theoretical work on common agency theory to a cooperative framework, and some issues concerning control in CMAs are discussed in the first sections of this paper. A few examples of CMAs are examined, and some of the patterns from these experiences are considered as possible constraints to achieving a market development objective with a CMA form of cooperation.

Multiple-Principal/Single-Agent

A "multiple principals-single agent" relationship is a distinguishing feature of CMAs. It is a variant of the standard principal-agent model. Before justifying the notion of multiple principals in a CMA, while a direct membership producer cooperative has a single principal, one should be aware that the principal-agent model has not been widely adopted in theoretical work on cooperatives.   Some agricultural economists have used the related theory of agency, where both parties to a cooperative arrangement, members and management, are agents who functionally relate to one another in a nexus of contracts.   The economists, Holmstrom and Tirole, point out that agency theorists who emphasize capital structure interpretations of the firm tend to follow the convention of viewing both parties to a contract as agents.[1]   In this same vein, cooperative theorists typically refer to the board of directors as the "representatives of the residual claimants", rather than use the term principal.[2]

An advantage to using a principal-agent model is that it clarifies the meaning of control, when it is effective or ineffective, as well as the nature and direction of incentives in a cooperative.   In their survey article, Holmstrom and Tirole discuss some of the uses of principal-agent relationships in the theory of the firm.   They note that the weakness of this approach is that directors, supposedly carrying out the control responsibilities of a principal, often have close ties with management and opportunities for collusion.[3]   However, when we consider the evidence of what kind of directors there are in cooperatives versus investor-owned firms, as examined by Caswell, this objection to the principal-agent model is invalid or at least insignificant when applied to a cooperative form of business.[4]

In recent theoretical work on the concept of common agency, Bernheim and Whinston use the term "multiple-principal/single-agent".[5]   Their major application is to the wholesale trade, where manufacturers contract with merchandise agents or brokers to market their products to retailers.   One of their requirements for common agency is that the principals do not cooperate in selecting and in establishing a fee structure for their common agent, nor do they communicate about prices.   In their theory, when such cooperation exists, the arrangement becomes the standard principal-agent model.

The opportunities provided by the Capper-Volstead Act for cooperatives to actually establish a formal CMA is a feature that Bernheim and Whinston do not address.   Whether farmer cooperative CMAs fit their theory or provide a rationale for a revised theory of common agency, is a complicated question that cannot be adequately considered at this time, but their work is supportive of a couple points.   They are the first agency theorists to develop the idea and conditions for multiple principals-agent relationships, which provide direction in exploring some overlooked differences between a CMA and a cooperative with direct producer membership.   Specifically, this distinction clarifies some of the governance and performance contrasts when evaluating a merger alternative versus a CMA.

In passing, the article by Bernheim and Whinston, "Common Marketing Agency as a Device for Facilitating Collusion" in the *Rand Journal of*

Economics, Spring 1985, is particularly interesting in light of the recent trend by Del Monte and Heinz to replace their sales network with contractual arrangements with broker agencies.

The assumption of multiple principals raises a question as to why all cooperatives are not viewed as having more than a single principal-agent relationship. Multiplicity and subsequent conflict exist among members or on the board of directors of most cooperatives. However, the distinction hinges on the fact that cooperatives, unlike producers, have developed distinctive expertise, asset specificity, and intellectual property over many years. Each cooperative has a unique set of these factors or resources that it would like to continue to augment and to protect. In addition, the assumption of a single principal is that members have combined in a cooperative, whereas a CMA is far from a full combination. The members of each cooperative in a CMA are probably reluctant to merge, and this reluctance is another factor in addition to specialized resources that renders multiple principals as a defining characteristic.

Control

The meaning of control and the concerns over failure to exercise it in cooperatives and business organizations has occupied much attention in academic work and in the media with regard to executive compensation. Much earlier attention followed in the wake of Berle and Means work from the 1930s on the historical transition from owner-manager businesses to the stockholders-manager organized corporation. They argued that not only has ownership been separated from management, which those of us involved with cooperatives see as an alternative with certain advantages, but the stockholding owners have lost control. Their adopted phrase "separation of ownership and control" is unfortunate because it has resulted in the notion of a principal's responsibility to control as an activity that management can appropriate. A more accurate expression is that control is lacking.

Control over the direction of a business entity is not what managers do, but is a concept that has to do with monitoring managerial performance and directing a firm's course of action, unless one wanted to introduce the idea of managers having self-control. Harold Demsetz noted this distinction when he used the term "control vacuum" to discuss the separation of ownership and control, although he thinks its a non-issue.[6] Fama and Jensen also grasped the issue of control as having at least elements that only have meaning with regard to the principal by making the distinction between "decision control" and "decision management." However, in his earlier writings, Fama follows the convention of managers having seized control in the modern corporation.[7] In the context of farmer cooperatives this important distinction has been made by Andrew Condon.[8]

The responsibility of cooperative membership for decision control can also be divided between governance control and performance control. Much of the emphasis by ACS is on the former, reflecting the concerns about maintaining the integrity of a cooperative and about professional management moving a cooperative into business opportunities that provide high returns but do not enhance the value of member products. Successful enhancement of the value of member products through a long-run market development program is the case where both governance and performance control are working. Members capture part of this gain in increased value of their production resources, whereas not all increases in a cooperative's earnings would have such an impact. The latter might reflect a situation where governance control was lacking.

A merger condition maintains the same principal-agent structure as in the pre-merger situation, whereas a CMA creates a multiple principal structure. A multiple principal structure renders both governance and performance control especially difficult. Each member of a CMA wants to garner the most benefit and suspects that other members are either receiving more than their fair share or are not sharing enough of the costs. A large part of control in CMAs involves governance in establishing rules for how benefits and costs will be equitably shared. A similar point about the problem of control has been made in the context of hierarchial or layered structures of cooperatives.[9]

When cooperatives form a CMA, performance control issues are less salient than those of governance, but they are possibly more serious. In contrast, governance is the major control challenge when cooperatives confront a decision to merge or not. In fact, when merger discussions break-off it is often due to uncertainties about governance control, and rarely is it a question about a proposed merger creating business inefficiency. In addition, a CMA alternative provides more assurance that governance will not be a problem since the participating cooperatives remain in existence. In other words, the choice between merger or CMA is to some extent a trade-off between governance and performance control.

Ineffective performance control leads to CMAs, that no matter how structured, do not achieve their potential in either price negotiation, operating efficiency, or if attempted, joint market development. The latter is the most difficult to monitor and evaluate because it involves a lot of intangibles, such as developing innovative product modifications and establishing a strong brand. In contrast, weaknesses in governance control are more easily detected than are failures to achieve marketing gains or to develop new sources of demand for member products.

Examples of CMAs

Most relatively large and financially solvent cooperatives are not prone to seeking merger partners. When constraints or inefficiencies are encountered, particularly those of suboptimal economies of size, cooperatives will seek some type of strategic alliance—CMA or other arrangement. This situation is often accompanied by cooperatives wanting to maintain marketing programs for the core of their member products, and are able to identify either a special function or market segment that is subject to more economies of size than they individually can achieve. By-products marketing is an example. Cooperatives will not focus their resources on by-products because management is evaluated predominantly on how it enhances prices and develops marketing programs for the primary product of its members. Midwest Agri-Commodities, Inc. is a case in-point. It merchandises beet pulp and molasses for three sugar beet processing cooperatives.

Assuming that the relevant range of the average cost function for marketing beet sugar by-products is downward sloping, a CMA reduces costs in comparison to the total of each cooperative's cost for the same volume of by-products sold. The marketing of beet sugar may have a similar cost structure, but it would probably be more difficult to consolidate this operation. If a CMA were to take responsibility for the total output of sugar as well, it raises the question of why not merge. The latter involves elimination of each original cooperative's identity, which grower-members often oppose for reasons of governance control. A CMA may also involve complex situations of having to equitably share a mix of specialized assets, business expertise, and intellectual property of each cooperative.

The by-products advantage for setting-up CMAs is also demonstrated by the contrast between soybean and cottonseed processing. In 1963 several cooperative soybean and cottonseed processors established a CMA called Soy-Cot Sales, Inc. The soybean side was never able to effectively coordinate its marketing in this organization, but by the early 1980's, competitive pressures had reached a point that industry participants either achieved a certain economies of size threshold or exited the industry. A merging of soybean processing cooperatives or divisions into Ag Processing, Inc. was the outcome. With the cottonseed processors, Soy-Cot continues to operate as a successful CMA, handling only cottonseed oil for its members.

Mentioning cottonseed immediately brings up the seemingly counter-example of a cotton CMA, AMCOT, Inc. However, the by-products advantage is just that, an advantage and not a necessary condition for a successful CMA. In regard to AMCOT, it should be pointed that CMAs having cooperatives that do not directly compete with one another, reduce the extent of members' concern about having the same marketing opportunities for their products as other members receive. It is misleading to think of the AMCOT members' cotton as perfect substitutes, as would generally be true in the case of grain cooperatives. In fact, one of the reasons AMCOT works as well as it does is that each of its four members handle significantly different growths. AMCOT is to a large extent a multicommodity cooperative. In order to provide effective sales representation in world cotton markets, an agent needs to represent a wide variety of cotton. This situation is perfect for cotton cooperatives, limited to the particular kinds of growths in their member region, they have achieved a threshold economies of size by establishing a CMA.

Another segmented market situation that lends itself to the establishment of CMAs by cooperatives is when there exists a need to handle non-member sales as a separate operation. This condition is present in the cattle artificial insemination industry (AI). The AI cooperatives are involved with producing an input and providing substantial technical service to their members for optimizing its application. Production above member needs is available for non-member sales without providing the same amount of technical services. Non-member sales, including exports, is a separable operation. In terms of separability, it is akin to the handling of by-products.

Three different groups of AI cooperatives have established CMAs to handle non-member sales and exports. They have also collectively contracted a fourth CMA, known as World Wide Sires, to handle exports to Europe, Africa, and Asia. By specializing in non-members sales, these CMAs are organized as partnerships without incorporating as cooperatives. The key point is that the value-added components of this business are controlled in a more localized or regional structure of individual cooperatives, while activities where control is less of a concern, are the components that are organized into CMAs.

There are examples of participation in CMAs by cooperatives with highly developed marketing programs and established brands. Sun-Diamond is the most well-known of these. Much of its coordination responsibilities involve public policy and relations. Sun-Diamond's marketing responsibilities focus on activities and functions that are subject to economies of size. It does order entry, coordinates distribution, handles accounts payable and receivable, manages a network of foreign sales representatives, and handles export documentation. Member cooperatives are responsible for market development—advertising, product improvements and developments, selection of product lines for export, sales objectives, and pricing. These latter activities may also be subject to economies of size, however, they comprise the key economic assets of each cooperative.

Sun-Diamond's acquisition of a proprietary dried fruit and nut company, Sunland, has brought to the fore the issue of coordination and compromise in the use of brands. Sunland's brands were vacated and its product lines have been marketed under various trademarks of the members. Sun-Diamond is responsible for managing the Sunland division and selecting which member brands are to be used. This area of CMA operations has been one of the most contentious for members. While it is different from a situation of a CMA managing a members' brands and the development of its markets, it does involve decisions over the use of brands that are not under the complete control of the members.

An example of a CMA that is organized to handle the brands of member cooperatives for specialized packaging for the institutional market is provided by Cooperating Brands, Inc. It was established by six fruit processing cooperatives for the purpose of managing a special institutional juice pack. A small frozen juice cup is a required type of packaging for a certain segment of the institutional market. None of the cooperatives wanted to invest in this type of packaging, and preferred co-packing arrangements. However, there are significant economic advantages to having both large volume co-packing arrangements and a comprehensive line of juice varieties when selling to institutional buyers. Cooperating Brands was established in 1985 to manage co-packing and sales for this special institutional pack. Opportunities for special packs of different kinds of fruit snacks have also developed. Annual sales have recently averaged about $20 million for the West Coast market.

Cooperating Brands has a trademark licensing agreement with each member cooperative. In a co-packing arrangement, the packing plant does not take title, and they explicitly have no rights to use the brands on their own products. A confidentiality agreement restricts the packer or bottler from making any use or release of the proprietary juice formulations. Recently member cooperatives have, for most of their products, been able to prevent this problem by shipping pre-formulated concentrates to the bottlers.

These contractual arrangements indicate the extent of proprietary information, and hence, specific capabilities that value-added processing cooperatives possess. Juice marketing cooperatives are highly successful and competitive with one another. Cooperating Brands, Inc. provides a means for gaining access to a specialized and segmented market, while protecting their distinctive competence from duplication by competitors. Cooperatives that have developed intellectual property and special technical expertise in food processing and marketing, will use CMAs to achieve economies for better access to particular market segments, but they will be contractually individualized and far removed from a product pooling concept.

Earlier comments about Midwest Agri-Commodities, AMCOT, and Soy-Cot provide examples of CMAs that do not involve major brand-name products and operate in price competitive commodity markets. In fact, most CMAs have been of this type. The milk marketing CMAs, Central Milk Processing Cooperative and the Regional Cooperative Marketing Association have been able to help stabilize prices and have been able to achieve over-order premiums.

In apple marketing in the Northwest, federated marketing for cooperative packing houses goes back as far as 1922, and persists today with Wenoka Sales. It markets directly to the grocery chains for seven members, and, although each has its own packing house brand, such trademarks do not have licensing value. Wenoka has found that recently these buyers have been more receptive to being able to deal without broker intermediaries. The fact that the grocery chains want to buy direct is perhaps an indication of their fear of the subtle collusive process that a broker can provide, as examined by Bernheim and Whinston. In any event, Wenoka's members have more control over pricing coordination and negotiating strength, in contrast to delegating these functions to an outside broker.

The challenge of maintaining product commitment from the members of a CMA, particularly when operating in commodity markets, often raises the issue of marketing agreements. This issue is not addressed in this paper, but it is one that merits more research and economic analysis. Many commodity oriented CMAs have failed after operating for several years. A contributing factor to their demise was a lessening of product commitment by some members. This has occurred in grains with Producers Export Company and Farmers Export Company, as well as in dried beans, with the recent termination of Valley Marketing, Inc.

Without exploring all the reasons for the demise of these CMAs, it is worth observing that these examples exhibit the segmentation pattern that has been discussed above. A group of cooperatives identify a particular marketing function or a market segment that they are willing to delegate to a CMA. Valley Marketing consisted of local grain cooperatives that handled a relatively small volume of beans from some of their members and a couple of predominantly specialized dried bean cooperatives. It was inefficient for the grain cooperatives to each market a small volume, so a CMA worked well for them. However, their interests eventually came into conflict when the two local dried bean cooperatives wanted to build their own processing facilities for frozen beans. This placed pressure on Valley Marketing to procure non-member beans to maintain customer accounts. The two largest members opposed the idea of non-member business and dropped out of the organization, leaving it with an unsustainable level of volume.

The case of the grain marketing interregionals reflected the segmentation pattern around the idea of exporting. They reasoned that grain exporting, particularly in operating port elevators, was subject to more economies of size than they could individually attain. Hence, the regional cooperative members handled domestic marketing, while the interregional specialized in exporting. The problem was that grain markets cannot be segmented in this way. A successful CMA in grain marketing would require responsibility for both export and domestic marketing decisions. The desire to delegate special activities to a CMA can work in many contexts, as discussed above, but was a situation destined for failure in grain marketing.

Constraints to the Use of CMAs for Market Development

Many aspects of market development and value-added activities are subject to substantial economies of size. It is particularly difficult to have adequate financial resources to establish major brands and to gain access to shelf space in the major super market chains throughout the world. CMAs are viewed as an option for achieving this type of market development, with the various products of different cooperatives being strategically positioned into a family of products approach.

Most examples of CMAs cover by-products or other situations of a segmentation pattern when involving products that have benefitted from highly developed marketing programs of their direct producer membership cooperatives. The strategy of forming a CMA to pursue the market development objective is of current interest. Robert Cropp has pointed out that within the dairy cooperative community there is debate over whether or not to include highly developed marketing programs of strong brand-name products of some participants in a CMA, or to have them operate a separate program on the outside.[10] In the latter case, they would commit some target volume of raw product to a CMA.

Although examples of forming a CMA that would pool the resources of members into pursuing the market development objective are few in number, Norbest stands out as a potential model. However, it is important to understand that its accomplishments are not a recent development. Formed in 1929 to handle turkey sales for twenty cooperative packing associations, Norbest was an innovator in using the branded concept for marketing fresh meat, and up until recently, was the largest marketer of turkeys in the world.

In examining aspects of common agency theory and empirical attributes of CMAs above, a few constraints emerge as to the potential of accomplishing a market development objective. A multiple-principal/agent form of organization is likely to encounter more demanding requirements for exercising effective control than the standard principal/agent form of cooperative. It was pointed out that a CMA is often an alternative to a merger of cooperatives. Members of a CMA want to maintain a degree of governance control that a merger would seem to jeopardize. By maintaining their original cooperative associations and entering a CMA arrangement under restrictive terms with regard to their financial risk exposure, the requirements for governance control can easily be met.

Having avoided a merger, the principals of a CMA would need to compensate for a potential loss of efficiencies by implementing an effective system of performance control. One source of potential inefficiency is that managers of some CMAs have required substantial political skills to keep the organization together, and this requirement may detract from business performance.

The demands of performance control in a multiple principal agency are exacerbated when the objectives are complex. Of the four general objectives — (1) market power, (2) economies of size, (3) information sharing, and (4) market development; the latter involves more intangibles and less clear cut direction. The performance factors in developing markets for member products are more difficult for the principals to monitor than are market power objectives in price negotiations or efficiencies in operations. Hence, performance control tends to be less effective in CMAs, and while being adequate for accomplishing clearly definable and attainable objectives, is likely to be inadequate for directing management to achieve the kind of high level of innovation that market development requires.

Summary

In summary, the issues explored in this paper started with consideration of possible extensions of cooperative theory to accommodate unique aspects of a CMA. A possible direction is common agency theory. At this stage most cooperative theorists have yet to accept and apply the principal-agent model, let alone, the notion of multiple-principal/single-agent.

A second area that closely relates to both theory and practice is the meaning of control. If viewed in the context of either multiple or single principal, control is not a tug of war between directors and management. Rather, control is a responsibility of directors, and if they abdicate it, a control vacuum is created.

A third point is that control tends to be regarded exclusively as a governance concept, while the performance dimension is either passively recognized or even neglected. The exercise of performance control is often inadequate in cooperatives, especially in CMAs.

When examining examples of CMAs, a pattern of segmentation can be identified. CMAs tend to be responsible for a special function or market that can be economically segregated from the marketing operations of its member cooperatives.

Lastly, given the constraints on effective performance control, the objective of market development is likely to be more difficult to accomplish in the multiple principal setting of a CMA than in a single principal-agent cooperative.

Notes

1. B.R. Holmstrom and J. Tirole, The Theory of the Firm, *Handbook of Industrial Organization*, 2nd edition, 1990, p. 79-80.

2. John M. Staatz, Recent Developments in the Theory of Agricultural Cooperation, *Journal of Agricultural Cooperation*, V.2, 1987, p. 85.

3. Holmstrom and Tirole, p. 87-88.

4. Julie A. Caswell, The Cooperative-Corporate Interface: Interfirm Contact Through Membership on Boards of Directors, *Journal of Agricultural Cooperation*, V.4, 1989.

5. B. D. Bernheim and M.D. Whinston, Common Marketing Agency as a Device for Facilitating Collusion, *The Rand Journal of Economics*, V. 16, No.2, Summer 1985. Also see their article. Common Agency, *Econometrica*, V. 54, No. 4, July 1986.

6. H. Demsetz, The Structure of Ownership and the Theory of the Firm, *Journal of Law & Economics*, V. 26, June 1983, p. 387.

7. E.F. Fama and M.C. Jensen, Separation of Ownership and Control, Ibid, p. 303-308. E.F. Fama, Agency Problems and the Theory of the Firm, *Journal of Political Economy*, V.88, No.2, 1980.

8. A. Condon, The Methodology and Requirements of a Theory of Modern Cooperative Enterprise, *Cooperative Theory: New Approaches*, USDA, ACS Service Report No. 18, 1987, p. 24-25.

9. Ibid, p. 27.

10. R.A. Cropp, Co-op Opportunity for Agency-in-Common Approach to Manufactured Dairy Products, *Farmer Cooperatives*, April 1992, p. 17.

References

Bernheim, B.D. and M.D. Whinston. 1985. Common Marketing Agency as a Device for Facilitating Collusion. *The Rand Journal of Economics* 16(4):269-281.

___. 1986. Common Agency. *Econometrica* 54(4):923-942.

Caswell, Julie. 1989. The Cooperative-Corporate Interface: Interfirm Contact Through Membership on Boards of Directors. *Journal of Agricultural Cooperation* 4:20-27.

Condon, Andrew. 1987. The Methodology and Requirements of a Theory of Modern Cooperative Enterprise. In Jeffrey S. Royer, ed., *Cooperative Theory: New Approaches*.

Cropp, R.A. 1992. Co-op Opportunity for Agency-in-Common Approach to Manufactured Dairy Products. *Farmer Cooperatives* April: 12-17.

Demsetz, Harold. 1983. The Structure of Ownership and the Theory of the Firm. *Journal of Law & Economics* 26: 375-393.

Fama, E.F. and M.C. Jensen. 1983. Separation of Ownership and Control. *Journal of Law & Economics* 26: 301-325.

Fama, E.F. 1980. Agency Problems and the Theory of the Firm. *Journal of Political Economy* 88(2):288-307.

Holmstrom, B.R. and J. Tirole. 1990. The Theory of the Firm. In *Handbook of Industrial Organization*. R. Schmalensee and R. Willig, editors, New York: Elsevier Science Publishers B.V.

Staatz, John M.. 1987. Recent Developments in the Theory of Agricultural Cooperation. *Journal of Agricultural Cooperation* 2: 74-95.

8

The Dairy Marketing Initiative of Upper Midwestern Cooperatives

William D. Dobson and Robert A. Cropp

Spurred by producer discontent over $10 per hundredweight manufacturing milk prices and bleak prospects for higher government support prices, 14 dairy cooperatives in the Upper Midwest launched a Dairy Marketing Initiative (DMI) in 1991. Upper Midwestern Dairy Cooperatives and other producer-controlled organizations will use the DMI to work together to increase milk prices, strengthen their bargaining power in the marketplace, and maintain the leading role of the Upper Midwest in the dairy industry. The DMI includes plans for several joint ventures or Common Marketing Agencies (CMAs) for dairy products, including fluid milk, cheese, other manufactured dairy products, and dairy exports. It also calls for efforts to reduce milk assembly and dairy plant costs. These comments represent an early evaluation of the relatively young DMI. Specifically, these remarks consist of a discussion of (1) the origins of the DMI, (2) progress to date on implementing the DMI, emphasizing the CMAs, and (3) what has been learned about the potential effectiveness of CMAs as tools for improving cooperative and industry performance. As will be apparent, the challenges facing dairy cooperatives using CMAs and other strategic alliances are great, but it is also clear the Upper Midwestern dairy cooperatives are doing a number of things correctly.

A CMA is a group of marketing cooperatives that market products under a common agreement. Under a CMA, member cooperatives maintain operational independence while hoping to capitalize on the earnings and market power of a larger organization. A pure CMA is a federated cooperative whose sole responsibility is to serve as a marketing agent for its members. It does not physically handle products, and generally does not take title to them. However, some cooperatives form CMAs that consist of federated marketing cooperatives that take title to members' products and have major responsibility for handling, pricing, and marketing the products. Those launching the DMI are likely to be eclectic, and use both types of CMAs or something in between. An association of milk producers qualifying as a cooperative under the Capper Volstead Act of 1922 can employ a CMA.

Origins of the Dairy Marketing Initiative

The DMI has its origins in the low milk prices, increased price variability, and concerns about imbalances in market power that surfaced in 1990-91. As is well known to many in this group, the Minnesota-Wisconsin (MW) manufacturing milk price fell to about $10 per hundredweight during December 1990 and early 1991. This development impressed upon Upper Midwestern cooperatives that the existing federal price support regime under which dairy price supports had fallen 25% (50% in real terms) from their 1981 peak to the current $10.10 level carried with it times of low milk prices and greater price variability. The variability and downside price risk were underscored by the drop in the MW price from $13.94 in January 1990 to $10.19 in December 1990, a 27% decline. This drop in prices had obvious effects on dairy farmers' cash flows. Operators of manufacturing milk plants also found it difficult to manage product inventories. Many experienced substantial losses from reduced inventory values when cheese prices plummeted during the fall of 1990. In a classic understatement, a Land O'Lakes official said, "Companies make mistakes when milk prices are as variable as in 1990-91."

Kraft's dominance in branded cheese markets also was part of the economic environment. A Land O'Lakes official pointed out that Kraft produces only about 20% of its own cheese, while 80% of its cheese is produced in non-Kraft plants--including many cooperative plants in the Upper Midwest (Cheese Reporter, March 6, 1992). It was claimed that milk producers supplying these plants didn't reap the benefits of the Kraft name and share in Kraft's profits. In early 1991, many milk producers also were angered by Kraft's decision to maintain high cheese prices despite falling milk prices, which curtailed cheese sales. Thus, the cooperatives were encouraged to see whether they could obtain some of the profit margin that Kraft earned on cheese produced by the cooperatives through joint action. Ironically, Kraft's pricing strategy hurt the company in the supermarket since it lost market share to private label cheese products the prices of which were cut rapidly when raw product costs declined in 1990-91 (Deveny, 1992).

More generally, there were concerns about the growing market power of investor-owned dairy companies and the apparent lack of feasibility of mergers

TABLE 8.1 Cooperatives Participating in Dairy Marketing Initiative, 1992.

A.G. Cooperative, Arcadia, WI
Alto Dairy Cooperative, Waupun, WI
AMPI, Morning Glory Farms, Shawano, WI
AMPI, North Central Region, New Ulm, MN
Cass Clay Cooperative, Fargo, ND
First District Association, Litchfield, MN
Golden Guernsey Dairy, Milwaukee, WI
Land O'Lakes, Inc., Arden Hills, MN, and
Lake to Lake Milk Producers, Kiel, WI
Mid-America Dairymen, Inc., Springfield, MO
Milwaukee Cooperative Milk Producers, Brookfield, WI
Midwest Dairymen, Rockford, IL
National Farmers Organization, Holman, WI
Tri-State Milk Producers, West Salem, WI
Swiss Valley Farms, Davenport, IA
Wisconsin Dairies, Baraboo, WI

---

of cooperatives to offset this power imbalance. Don Storhoff, CEO of Wisconsin Dairies, made the following comment, variations of which were heard frequently, "There are too many structural and policy differences to allow mergers right now. I don't see any major mergers coming up soon (McNair, 1991)."

Individually and in group meetings, Upper Midwestern dairy cooperatives considered options for dealing with the low and variable milk prices in late 1990 and early 1991. Efforts, which still continue, were made to lobby for higher dairy price supports. However, because of poor prospects for higher supports attention quickly turned to a parallel effort that involved joint work by the cooperatives.   With coordinating help from the Wisconsin Federation of Cooperatives, the cooperatives developed the DMI. Initially, at least, the DMI is structured to include cooperatives in North Dakota, South Dakota, Minnesota, Iowa, Wisconsin and Illinois. Other "farmer-controlled" organizations that are not dairy cooperatives can be included in the DMI. As shown in Table 8.1, organizations involved in the DMI include Land O'Lakes, Wisconsin Dairies, Mid-America Dairymen, and Associated Milk Producers, Inc., some of the largest cooperatives in the U.S.

As noted earlier, the DMI includes plans for CMAs for fluid milk, cheese, dry whey products, and exported dairy products. It also calls for investigations of efficiencies that might be gained from plant consolidations, more efficient milk pickup arrangements, and ways to reduce farmers' health insurance costs. The emphasis placed on these items in the planning stage was as follows:

Item	Percent of Total Budget
Marketing Fluid Milk	5-10%
Marketing of Whey Products, Cheese, and other Manufactured Dairy Products	35-40%
Dairy Exporting	10-15%
Transportation of Milk	10-15%
Dairy Plant Efficiency	15-20%
Health Care for Farmers	10-15%

Thus, early plans called for the largest amount of resources to be targeted at marketing efforts regarding whey products, cheese and other manufactured dairy products. This effort envisioned completing plans and negotiations for a whey products CMA and a study of the feasibility of joint marketing of cheese and other manufactured dairy products. Efforts regarding dairy plant efficiency, the item with the next largest budget, would involve analyzing future dairy plant capacity needs in the Upper Midwest and opportunities for joint action to improve plant capacity and technology utilization.

Progress to Date in Implementing the Dairy Marketing Initiative

Progress must ultimately be assessed by comparing results achieved to the goals for the initiative. As indicated earlier, the overall goal of the DMI is for Upper Midwest Dairy cooperatives and other producer-controlled organizations to work together to increase milk prices, increase profits of their members, strengthen their bargaining power in the market place, and maintain the leading role of the Upper Midwest in the dairy industry. Objectives for selected individual items in the DMI are as follows:

1. Fluid Milk: Increase and stabilize milk prices by assisting in establishing and strengthening fluid milk CMAs.
2. Whey and Other Manufactured Dairy Products: Increase and stabilize prices for whey and other manufactured dairy products by identifying and implementing joint actions by DMI members.
3. Export Marketing: Thoroughly analyze opportunities for joint export marketing by Upper Midwest dairy organizations.
4. Transportation: Reduce raw milk and milk product hauling costs by cordinating transportation services among dairy cooperatives and dairy companies.
5. Dairy Plant Efficiencies: Reduce the costs of processing dairy products

by consolidating plant operations where appropriate, improving plant technology, and reducing the amount of time plants are operating below capacity.

It is too early to tell how effective the joint activities and CMAs envisioned as part of the DMI will be for achieving these objectives. Proposed priorities and work plans were placed before a farmer-cooperative director policy committee that decides priorities and makes financing recommendations. And in April 1992, the policy committee established the following three priorities for 1992:

- Improve Upper Midwest fluid milk prices by helping DMI members to establish and strengthen over-order pricing programs.
- Promote and support joint action by DMI members on R&D and processing and marketing of whey and other manufactured dairy products.
- Promote and support a pilot, joint export marketing project by interested DMI members.

Observations can be made about likely progress toward reaching objectives associated with these priority areas.

Promising Early Signs

The organizational and planning efforts have been noteworthy for their lack of extravagant claims for the DMI and for recognition on the part of the cooperatives of problems associated with CMAs. For example, early discussions suggested that a $.30 to $.40 per hundredweight increase in farm milk prices would be the most that could be obtained from a successful effort. Secondly, there have been few demands by managers and directors of the cooperatives for a "quick fix" for low returns in dairying. Instead efforts have been put in place to determine which CMAs are likely to be most successful over the longer-run. This situation has been accompanied by support for market research by the participating cooperatives and from three other sources. First, the Wisconsin Department of Agriculture, Trade and Consumer Protection provided a grant of $50,000 for financing market analysis, planning, and legal assistance involved in developing CMAs. Secondly, the Agricultural Technology and Family Farm Institute at the University of Wisconsin has provided about $40,000 for research on the feasibility of a CMA for barrel cheese. Finally, the Wisconsin Milk Marketing Board (WMMB) has offered to provide support for research ventures involving DMI members seeking to expand sales of dried whey and other manufactured dairy products.

The cooperatives entered the initiative recognizing the difficulties associated with asking cooperatives that have developed well-known brands or differentiated their products in other ways to share the returns from the differentiation measures with other cooperatives through a CMA. This point is important because several of the cooperatives—e.g., Land O'Lakes, Wisconsin Dairies, and Mid-America Dairymen—have differentiated many of their products. The famous Land O'Lakes brand for butter comes to mind in this connection. It is generally understood that differentiated products sold by a cooperative will either have to be omitted from a CMA or that cooperatives will require compensation before they will share returns from product differentiation under a CMA. Fresh in mind too was an effort entered into in 1986 involving some participants in the DMI that was aimed at joint marketing of dry whey products. That effort was abandoned when milk prices increased. Presumably the history of this effort will prevent early abandonment of similar efforts under the present DMI.

Mechanisms Already in Place

Common Marketing Agencies are already in place for fluid milk in the Upper Midwest. The most prominent example is the Central Milk Producers Cooperative (CMPC), a group of 13 cooperatives and the National Farmers Organization that supplies handlers in the Chicago Regional Federal order market. CMPC, which includes cooperatives that participate in the DMI, has successfully negotiated some of the highest Class I price premiums in the nation for the Chicago and Milwaukee markets. CMPC reports it has collected an average over order premium of $2.50 per hundredweight on Class I milk and $1.50 per hundredweight on Class II milk over the past three years (Cheese Reporter, March 13, 1992). The premiums reflect CMPC's 90% share of the market supply for the Chicago Regional milk order and compensation for the milk assembly and balancing services provided by the organization. Nevertheless, the 10% of the milk not under CMPC's control has been sufficient to cause undercutting of milk prices and required the organization to give competitive credits to milk buyers. There is an additional complication. Not all DMI cooperatives with fluid milk sales are members of CMPC. The need for competitive credits has caused friction between DMI cooperatives that are members of CMPC and those not members of CMPC. Some cooperatives argue strongly for first solving problems generated by the CMPC membership profile before considering CMAs for other products. This explains why the DMI policy committee put first priority for 1992 on helping DMI members establish and strengthen over-order pricing programs for fluid milk.

The challenge for the DMI regarding fluid milk will include keeping CMPC member cooperatives working together, reducing the competitive credits required by handlers, and obtaining higher Class I premium prices for the Upper Midwest (Minneapolis-St. Paul) and Iowa federal order markets. Officials of cooperatives report that progress has been made recently in negotiating higher Class I price premiums for the Upper Midwest market, which historically has had substantially lower fluid milk price premiums than Chicago. While negotiating premium prices for fluid milk in somewhat localized fluid milk markets is a task for which a CMA is likely to be most effective, the challenges involved in successfully negotiating premium prices for a long period are still substantial.

A few cooperatives participating in the DMI have engaged in milk swapping to reduce milk hauling costs. In addition, research involving Associated Milk Producers, Inc. and a University of Wisconsin-Platteville faculty member has been carried out on computerized milk truck routing. These efforts should provide a useful foundation for additional work to obtain efficiencies in milk transportation.

Exporting

Several cooperatives participating in the DMI have experience in dairy exporting. Wisconsin Dairies, for example, through its Foremost Ingredient Group, exports lactose products (especially pharmaceutical lactose) to Latin America, the Pacific Rim, Europe, and Africa. The First District Association exports dry whey products to Pacific Rim markets. Mid-America Dairymen exports differentiated products—e.g., Sport Shake—and bulk, subsidized dairy commodities. Land O'Lakes makes small exports and is increasing its exporting experience by participating in trade fairs and trade missions. With a few exceptions, these are small, experimental exporting efforts. The personnel involved in dairy exporting also generally are spread thin and have major responsibilities in addition to exporting. Thus, the cooperatives have far less exporting experience and commitment to exporting than the New Zealand Dairy Board, Ireland's Dairy Board, DMV Campina, Kraft, Nestle, and Borden.

Preliminary committee work for the DMI suggests that (1) certain cooperatives could export more dairy products if they had the necessary milk and products, (2) the cooperatives could make additional export sales if they shared information on customer needs in export markets i.e., if they used a "market locator," (3) the cooperatives must devote more resources to exporting if they expect to be competitive with leading dairy exporters, and (4) there are opportunities for strategic alliances to increase dairy exports. On the last point, DMV Campina, a large Netherlands-based cooperative with a plant in La Crosse, Wisconsin, is eager to enter into strategic alliances with U.S. dairy cooperatives to export dairy products. DMV Campina regards the alliances as an effective way to spread its R&D and international marketing costs over more units of product. M.E. Franks would be willing to enter into contracts with the cooperatives to export bulk dairy commodities. Finally, Cooperative Business International of Washington, D.C. has offered its services to DMI for expanding exports of dairy products.

Progress on the CMA for Dry Whey Products

Investigations of the feasibility of using a CMA for dry whey products has proceeded farther than for other products. Cropp, Graf and Brown carried out market research to assess the feasibility of implementing a CMA for whey products. Findings from their survey of DMI cooperatives revealed the following points about the whey processing and marketing practices of the firms:

- The dry equivalent pounds of whey products processed by the DMI cooperatives totaled about 1.05 billion pounds per year, about 56% of the U.S. total.
- The cooperatives surveyed have whey marketing staffs that average 3.6 persons per cooperative, but which vary substantially in size from firm to firm.
- DMI member cooperatives market whey products through three main channels: other whey processors (11%), whey product manufacturing, including internal use (73%), and brokers (14%). The firms anticipate that the largest percentage increase in demand will be recorded for whey protein concentrate products with a protein content of 70-79%.

The DMI member cooperatives were questioned about current marketing practices and prospects for successful use of a CMA for dry whey. For certain survey items, the respondents assigned ratings ranging from 1 to 5, where 1 = Low and 5 = high. Average scores assigned by respondents for the items listed below are as follows:

Item	Average Score
Satisfaction with current marketing efforts	3.9
Perceived benefits of a commodity CMA	2.0
Willingness to consider current DMI cooperative members as CMA agent	2.7

These responses indicate relatively high levels of satisfaction with current marketing efforts and identify only moderate support for a CMA for whey products. The score of 2.0 for perceived benefits reflects the belief of some respondents that the DMI cooperative's share of the whey market is too small to have a significant effect on the price of whey. This score also reflects the belief that gaining agreement on which whey products to include in the CMA will be difficult to resolve. However, many respondents believe that a CMA could perform some important functions. The ability to share production and inventory information was mentioned as a valuable benefit by all respondents. As noted later, information sharing for dry whey products has commenced.

Increased price stability also was viewed as a potential benefit of a CMA for whey products.

Responses to a question about the acceptability of using a member cooperative as a marketing agent varied considerably. As part of this question, Land O'Lakes and Wisconsin Dairies were given as specific examples of firms that might serve as a marketing agent. The scores recorded for this item ranged from 1.0 to 4.5, with most scores being on the ends of the scale rather than the middle. The ambivalence about using a competitor as a marketing agent probably emerged in this response.

Respondents were asked what concession they would be willing to make to form a whey CMA. All were willing to include their commodity whey products, but several would not include specialty or blended whey products they produced in the CMA.

The market research of Cropp, Graf, and Brown provides figures that can be used to estimate price gains that might be achieved if a CMA for dry whey products based on supply withholding by cooperatives in the CMA was implemented. In addition, it was assumed that cooperatives receive zero return for dry whey products withheld from the market under the CMA. The analysis appearing below is carried out for two situations, which differ only regarding the share of national production in the hands of cooperatives participating in a CMA for dry whey products.

Situation No. 1. The dry whey production figures, production share in the hands of Upper Midwestern cooperatives participating in the CMA for dry whey, and the pricing objectives for this situation are as follows:

- Total U.S. dry whey production equals 1,175 million pounds per year.
- The share of U.S. dry whey production accounted for by Wisconsin, Minnesota, Iowa, and Illinois cooperatives participating in the CMA is 48%.
- The hypothetical pricing objective is to increase dry whey prices from $.20 to $.25 per pound (25%).

Situation No. 2. The dry whey production figures and the pricing objective under this situation are the same as under Situation No. 1. However, in this situation the dry whey produced by all U.S. cooperatives (83% of the national production) is assumed to be marketed under a CMA.

The results for four cases for situation No. 1 are shown in Table 2. Cases 1 through 4 were developed using different demand and supply elasticities to construct supply and demand functions and compute the national excess supply that would exist at the $.25 per pound target price for dry whey. The -.3 price elasticity of demand used in Case 1 was reported by Cropp, Graf and Brown. The -.37 price elasticity of demand is an estimate developed by Brandow for a residual group of dairy products, including milk powders. The supply

TABLE 8.2 Estimated Impact of Common Marketing Agencies on Dry Whey Prices.

Situation, Production and Price Items	Case 1	Case 2	Case 3	Case 4
Situation No. 1: (Upper Midwest Cooperatives in CMA)				
Price elasticity of demand used in case	-.3	-.37	-.37	-.37
Price elasticity of supply used in case	0.0	0.0	0.20	.40
National excess supply at $.25/lb	7.5%	9.2%	13.6%	17.5%
Reduction in dry whey marketings by WI, MN, IA and IL cooperatives needed to produce price of $.25/lb	15.6%	19.2%	28.3%	36.5%
Net price increase or decrease for WI, MN, IA and IL cooperatives resulting from joint activity				
• Proportion of dry whey marketed at $.25/lb x $.25	$ .211	$ .202	$ .179	$ .159
• Return from product withheld from market or dumped	0.0	0.0	0.0	0.0
• Net price	$ .211	$ .202	$ .179	$ .159
• Net price increase or decrease [(Net price - $.20/lb.)/$.20] x 100	5.5%	1.0%	-10.5%	-20.5%
Situation No. 2: (All U.S. Cooperatives in CMA)[a]				
National excess supply at $.25/lb	7.5%	9.2%	13.6%	17.5%
Reduction in dry whey marketings by all U.S. cooperatives needed to produce a price of $.25/lb	9.0%	11.1%	16.4%	21.1%
Net price increase or decrease for all U.S. cooperatives resulting from joint activity				
• Proportion of dry whey marketed at $.25/lb x $.25	$ .228	$ .222	$ .209	$ .197
• Return from product withheld from market or dumped	0	0	0	0
• Net price	$ .228	$ .222	$ .209	$ .197
• Net price increase or decrease [(Net price - $.20/lb.)/$.20] x 100	14.0%	11.0%	4.5%	-1.5%

[a] Price elasticities of demand and supply used for Cases 1-4 in Situation No. 2 are the same as under Situation No. 1.

elasticities are simply estimates used to test the sensitivity of the results to non zero supply response. Functions used in the study were positioned using 1,175 million pounds of dry whey and the $.20 per pound price as the equilibrium quantity and price, respectively.

Under Situation No. 1, the cooperatives in Wisconsin, Minnesota, Iowa, and Illinois were assumed to hold enough of their own production of dry whey products off the market to achieve the price objective of $.25 per pound. In Case 1 where the national excess supply is 7.5%, the cooperatives would be required to hold 15.6% of their production off the market to attain the price objective. A 5.5% increase in the net price of dry whey products is achieved through this action. This price increase reflected demand and supply functions developed with a price elasticity of demand for dry whey of -0.3 and the price elasticity of supply of zero.

Clearly, the results under the CMA are sensitive to the demand and supply elasticity estimates used for the analysis. In Case 2 where a slightly higher price elasticity of demand of -.37 and zero price elasticity of supply were used to construct the demand and supply functions, the cooperatives would hold about 19% of their production off the market and achieve a small (1.0%) increase in their net price for dry whey products. In Cases 3 and 4 where a -.37 price elasticity of demand was again used to construct the demand functions but higher .20 (Case 3) and .40 (Case 4) price elasticities of supply were employed to construct the supply functions, the reductions in the net price are larger. For Case 4 where the cooperatives must withhold 36.5% of their dry whey from the national market to achieve the price objective, they suffer a 20.5% reduction in the net price for dry whey.

The benefits from controlling a larger share of the marketings are obvious from the results for Situation No. 2. Under Situation No. 2, it is assumed that all U.S. cooperatives involved in marketing dry whey products participate in the CMA. The price objective and the supply and demand elasticities used to compute the functions needed to estimate the national excess supply and $.25 are the same as under Situation No. 1. The net price increase for Cases 1 and 2 under situation No. 2 are 14.0% and 11.0%, respectively. Only Case 4 which reflects the more elastic .4 supply elasticity produces a reduction in the net price for dry whey products.

The results for cases representing more elastic supply are noteworthy. The possibility exists for a somewhat elastic supply response for dry whey products since firms that now dump raw whey products will start processing whey to comply with antipollution regulations, especially if profits can be made from selling dry whey products under the higher prices that would be created by a CMA. The projected increase in cheese production also will generate whey as a byproduct that will be processed into a marketable dry whey product if there are profits to be gained from such processing.

The survey results noted earlier and results reported in Table 8.2 underscore the practical difficulties that Upper Midwestern dairy cooperatives would encounter if they attempted to implement a CMA based only on supply withholding by DMI members. The Free-Rider problem obviously would be substantial and possibilities exist that the net price for dry whey sold by participating cooperatives would be lower than under present marketing practices absent the CMA. The chances for increasing net prices for dry whey would be greater if the cooperatives could obtain a positive net return for dry whey products withheld from the market. An official of one cooperative suggested dumping the product overseas. The practical difficulties (e.g., retaliation, GATT violations, etc.) associated with this strategy were not discussed by the proponent. Positive net returns might be obtained by using the product withheld from the market for industrial uses that would not compete directly with regular commercial sales of dry whey. However, this option was not investigated.

What have the cooperatives decided to do regarding joint activity in the marketing of dry whey products? As a starting point, the cooperatives have decided to share information on the amount and location of their dry whey inventories and planned production of dry whey products. They noted that a similar mechanism now being used for nonfat dry milk by six cooperatives, including California cooperatives and Land O'Lakes, has been an effective bargaining aid for those cooperatives. Further, California dairy cooperatives have been actively investigating whether to use CMAs for nonfat dry milk, bulk butter and their Class I and Class II milk products. Because of difficulties encountered in developing acceptable CMAs and reservations the cooperatives had about using CMAs, the California cooperatives have initially organized an information exchange rather than a CMA. The DMI cooperatives believe information sharing will be a useful first step that will help them bargain more effectively with dry whey buyers and possibly lead to joint ventures and a CMA. Cropp, Graf, and Brown suggest that an interim joint venture involving only two or three cooperatives might be considerably more feasible to implement than a CMA.

Potential Usefulness of CMAs for Improving Performance

Only a few observations can be made about the usefulness of CMAs for improving cooperative and industry performance. The DMI does not appear to be generating bold strategic moves that would redress power imbalances between cooperatives and investor owned firms or take the cooperatives into potentially profitable new fields. In an effort to encourage the latter, a facilitator for the DMI effort suggested that the cooperatives join together and purchase M.E. Franks, the largest U.S. exporter of bulk dairy products, in order to become "big league" dairy exporters. This comment was dismissed with a chuckle as the cooperatives focussed energies on achieving more modest objectives.

Parenthetically, others apparently weren't chuckling at the idea. M.E. Franks was recently purchased by Ecoval, an investor-owned firm headquartered in Belgium.

The ways in which the modest efforts of the DMI might improve the performance of the cooperatives include the following:

1. The CMAS have been accompanied by useful market research that probably would not have been conducted otherwise. This market research will be useful for determining whether CMAs, joint ventures, or efforts by individual cooperatives will be the most promising strategies for increasing dairy farmer returns.

2. The act of getting together has helped the cooperatives to identify some useful actions. For example, as noted earlier the cooperatives involved in dairy exporting found that, as a group, they probably could increase export sales if a product locator was used to identify products that the cooperatives had available for export. Whether they will develop such a locator is unknown at this time. The efforts relating to a CMA for dry whey have led to information sharing on inventories and production plans for this product. The interest in pursuing additional reductions in raw milk hauling costs through milk swapping also was identified.

3. Information gathering efforts associated with the DMI have identified more clearly the difficult challenges facing producers of barrel cheese in the Upper Midwest. Partly because of raw product cost advantages, production of barrel cheese appears to be gravitating increasingly to the West Coast. The large processing plants built in the West to produce barrel cheese also possess important scale economies that give them advantages over the many plants in the Upper Midwest that compete fiercely for available supplies of milk for cheese production. More complete information on these developments may sharpen the strategic thinking of firms interested in retaining substantial production of barrel cheese in the Upper Midwest.

4. The publicity associated with the effort has attracted the interest of possible partners for joint efforts. DMV Campina, for example, has approached the cooperatives involved in the DMI to see if there is interest in forming strategic alliances for exporting. Similarly, M.E. Franks and Cooperative Business International, a subsidiary of the National Cooperative Business Association, have indicated a willingness to work with cooperatives in the DMI to expand dairy exports.

5. Discussion among DMI members has reinforced the idea that the CMPC needs to be strengthened and that a CMA is needed for fluid milk marketed in the Upper Midwest and Iowa federal order markets.

These developments could improve the performance of the cooperatives by making available to them more accurate information on market conditions, the market environment, possible partners for joint efforts, and the need to strengthen CMAs for fluid milk.

The challenges facing the cooperatives of course are many. Some cooperatives have found it difficult to devote enough time—especially time of senior managers—to investigating the feasibility of CMAs. While understandable, this has made it difficult to assess the depth of interest of certain cooperatives in CMAs.

Secondly, there are disagreements about whether to pursue a commodity approach or differentiated product approach in the CMAs. Some advocating a commodity approach argue that the best way to increase farmers' returns is to move a large amount of dairy solids into profitable markets. While recognizing that increased returns are obtainable through product differentiation, they argue that a product differentiation approach is likely to increase the returns of only a few cooperatives. Thus, they contend that product differentiation initiatives should be carried out by individual cooperatives or through joint ventures involving at most a few firms rather than a CMA.

Thirdly, there are major problems associated with attempting to increase profits of dairy farmers through use of CMAs for manufactured dairy products. Shares of production controlled by the Upper Midwestern cooperatives, while substantial, do not approach 100% for major manufactured dairy products. Thus, price enhancement efforts focussed on withholding supplies face the difficulties discussed in connection with the CMA for dry whey products. The options open to the Upper Midwestern cooperatives include (1) using CMAs for differentiated dairy products that are partially insulated from competitive actions of other firms, or (2) seeking to form strategic alliances with cooperatives outside the Upper Midwest to lessen free-rider problems and achieve other objectives.

It could be difficult to form strategic alliances necessary to deal with the free rider problem in view of the animosities that have developed among dairy cooperatives in different regions. However, the Standby Milk Pool established in 1967 represents a precedent for this type of initiative. The Standby Milk pool was a voluntary supply-management device (Cook, 1970). The basic principle of the Standby Pool was for cooperatives, which operated in markets where local supplies were about equal to Class I needs, to pay plants which were not regulated by a federal order and which had surplus Grade A milk supplies on hand, to make the milk available to deficit markets when and as needed. Under the Standby Pool, cooperatives in surplus milk production areas of the Upper Midwest shared in high priced deficit markets of the South. The presence of the device allowed producers in deficit milk markets to bargain more effectively for Class I price premiums. While the Standby Pool terminated in the mid 1980s and the effort it represented is not fully analogous to that required for developing a nationwide CMA for manufactured dairy products, it may be worth studying since it represents the general level of effort required to form multi-regional strategic marketing alliances in dairying. It also suggests that industry-sponsored supply control alliances are not impossible to develop.

Fourth, as this group appreciates, it will be difficult to measure the price gains achieved through use of CMAs for manufactured dairy products. Mechanisms for measuring results and the distribution of benefits among participants presumably will have to be established to satisfy the needs of participants for information. To date little attention has been given to this point.

Lastly, there can be opportunity costs associated with the joint action of the cooperatives. This is a point that Michael Porter of the Harvard Business School warns about. Porter has few good things to say about joint business efforts in any field, arguing that transitory alliances only give the illusion of productive activity and absorb precious management resources that individual firms might better use to upgrade their products and improve their competitiveness. Porter does concede that under certain circumstances cooperative research can prove beneficial. But, he says, the cooperative projects should be in areas of basic product and process research, not in subjects closely connected to a company's proprietary sources of advantage. One implication of Porter's warning is that cooperatives that have successfully differentiated their products might obtain higher returns from going it alone to upgrade and further differentiate their products. Porter's warnings concern issues that cooperatives in the DMI already seem to be acutely aware of. In particular, the cooperatives have shown little inclination to share resources closely associated with proprietary sources of advantage. Rather, their efforts appear to be devoted to areas where mutually beneficial gains might be possible. For example, joint efforts to obtain outside R&D information appears to be looked upon favorably by DMI cooperatives.

Summary

The DMI represents an ambitious joint effort entered into by Upper Midwestern cooperatives to deal with the low and variable milk prices that can accompany the present price support regime. It is frankly too early to tell how well the DMI will work. But there are some promising early signs. The expectations for the DMI have not been extravagant, there have been few demands for quick fixes, and market research is preceding action for individual CMAs. Previous work and mechanisms already in place—especially the CMAs for fluid milk—increase the chances that some of the CMAs will produce net benefits. Enhancing returns for major manufactured dairy products through use of CMAs appears to be the toughest challenge. While Porter's warnings about the pitfalls associated with joint efforts need to be kept in mind, they do not appear to pose a great risk to cooperatives associated with the DMI.

References

Brandow, G.E. 1961. Interrelationships Among Demands for Farm Products and Implications for the Control of Market Supply. *Pennsylvania Agricultural Experiment Station Bulletin 680* p. 59.

The Cheese Reporter. March 13, 1992. Congressional Report Examines Co-op Initiatives to Boost Milk Prices. p. 8.

The Cheese Reporter. March 6, 1992. LOL Dairy Foods Group Sees Branded, Value-Added Products as Keys In Future. p. 9.

Cook, H.L. 1970. The Standby Milk Pool—A New Strategic Bargaining Device. *American Journal of Agricultural Economics.* 52(1): 103-108.

Cropp, R., T. Graf, and M. Brown. 1992. The Feasibility of Joint Activities Among Dairy Cooperatives in the Processing and Marketing of Whey and Whey Products. *A Report to the Members of the Dairy Marketing Initiative.*

Cropp, R.A., E.V. Jesse, and W. Dobson. October 1991. Marketing Agencies in Common for Manufactured Dairy Products. *Marketing and Policy Briefing Paper No. 37.* Department of Agricultural Economics, UW-Madison.

Deveny, K. March 18, 1992. After Some Key Sales Strategies Go Sour, Kraft General Foods Gets Back to Basics. *Wall St. Journal.* p. B1.

McNair, J. January 17, 1991. Storhoff Says Co-ops Must Market Together. *Agri-View.* p. 3.

Porter, M.E. 1990. *The Competitive Advantage of Nations.* The Free Press, New York.

9

A Board Chairman's View of Requirements for Successful Common Marketing Agencies

Earl Giacolini

Sun-Diamond Growers of California is a common marketing agency serving five member cooperatives. I will describe the Sun-Diamond structure, explain how we conduct our business, look more specifically at Sun-Diamond's joint sales effort and describe the critical success factors for this partnership.

Let me begin by pointing out two basic prerequisites for the successful formation and operation of any CMA. First, because the legal aspects are subtle and complex, there is a need for counsel from attorneys who are experts in cooperative law. The second factor is timing. The time to form a CMA is when a cooperative is doing well, not when times are tough.

Sun-Diamond is a federated cooperative with five members and a wholly-owned corporate subsidiary, Sun-Land Products. Its members are Diamond Walnut Growers, Sun-Maid Growers of California, Sunsweet Growers, Valley Fig Growers, and Hazelnut Growers of Oregon. These cooperatives process and market raisins, walnuts, prunes, hazelnuts and dried figs under the Sun-Maid, Sunsweet, Diamond and Blue Ribbon brands—in the U.S. and 40 countries worldwide.

Figure 9.1 outlines the overall structure of the Sun-Diamond family. Each of the five member cooperatives is under the direction of its own board and management team. The five members own Sun-Diamond as a service organization. I should also explain that Sun-Diamond has 21 directors: five from each of the boards of the three major cooperatives; the chairman from Valley Fig; and the presidents of Diamond, Sun-Maid, Sunsweet, Valley Fig and Sun-Diamond.

Table 9.1 summarizes the 1991 financial performance for our combined associations. Combined sales of our dried fruit and nut products were $602.9

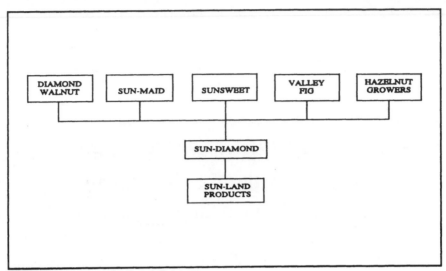

FIGURE 9.1 Organizational Structure

TABLE 9.1 1991 Financial Results ($ Million)

	Sales	Income	Member Proceeds
DIAMOND WALNUT	$171.1	$ 9.3	$112.5
SUN-MAID	179.0	8.6	80.6
SUNSWEET	185.5	18.9	85.5
VALLEY FIG	12.8	1.1	7.0
OREGON HAZELNUT	5.5	.5	2.8
SUN-LAND	49.0	3.8	3.7
Total	$602.9	$42.2	$292.1

million for the year. The table shows an income figure of $42.2 million, which is the amount earned above the field price from the processing and marketing of our commodities. The field price, which is the price paid by independent processors, is considered our commodity cost. Member proceeds is the total amount paid to growers for the farm-gate value of their commodities plus the income generated. For 1991 combined member proceeds were $292.1 million. While not shown on the chart, you may be interested to know that the growers who own these associations have invested equity capital in the businesses in the amount of $150 million. This demonstrates the confidence they have in their associations.

Table 9.2 outlines the history of Sun-Diamond, starting with its formation in

TABLE 9.2 History

- Sun-Diamond was founded in 1980 by Diamond, Sun-Maid, and Sunsweet. Valley Fig and Hazelnut joined later.
- Each had the #1 brand position but saw a need to improve sales effectiveness and reduce costs by joining together.
- Objectives
    - Increase sales and market shares.
    - Reduce sales and distribution costs.
    - Enhance political visibility and effectiveness.
- Centralization vs. Decentralization management.
    - 1980 — a centralized structure was adopted.
    - 1986 — restructured into a decentralized organization which provides each cooperative with direct control over its own affairs, yet offers advantages of partnership in Sun-Diamond.

1980 by Sun-Maid, Sunsweet, and Diamond. Valley Fig Growers and Hazelnut Growers joined several years later. Prospective partners in a CMA must identify a niche, just as we did in forming Sun-Diamond. It cannot have a broad, generic thrust that attempts to do everything for everybody. As the outline shows, in our case each of us had the number-one brand position in our product category. Each of us identified a need to strengthen the sales and marketing effort, enhance political visibility and effectiveness, and reduce costs by joining together. We're all branded, we're all 75-plus years old, we all have stable organizations, but we felt a need to look to the future. We also felt a common threat in that we compete against the Doles and the Del Montes of the world, large food companies with strong marketing leverage.

There are two ways of operating a CMA: centralized management or decentralized management. When we formed we adopted a centralized structure, meaning that the president and other officers of Sun-Diamond also served in those capacities for each of the member cooperatives. This seemed to work fine until 1985, when overpayments were made to Sun-Maid and Diamond growers and financial resources became strained. After much discussion and deliberation among the cooperatives, it was decided that we could have a more effective organization if we decentralized. Separate management teams were appointed for Sun-Diamond and each of the cooperatives. This revised format allows each cooperative to have more direct control over its own affairs, along with the advantages of the partnership in Sun-Diamond. Flexibility such as this is an essential ingredient for the long-term survival and success of a CMA.

Sun-Diamond acts as the agent for each cooperative, and the cooperatives reimburse Sun-Diamond for its operating costs. Each member cooperative retains title to its products and bears all business risks. Control over marketing, pricing, promotional programs and advertising, financial controls and financial reporting are handled at the member cooperative level. Although members work closely with Sun-Diamond, the member cooperative always has final say in these areas and ultimate responsibility.

The Sun-Diamond organization chart, Figure 9.2, reflects the primary functions of Sun-Diamond. Its key service areas are U.S. retail sales and international sales. Sun-Diamond also provides a complete logistics and distribution system which includes order entry, invoicing, transportation and warehousing.

In the financial area Sun-Diamond handles sales accounting, credit, cash collections, and cash and loan management. Sun-Diamond provides data processing services for our companies and has centralized systems that support sales and distribution activity. Sun-Diamond also provides computerized payroll support systems and other backup systems for the member cooperatives to help them with grower accounting and their other data processing needs.

Sun-Diamond's legal department provides an in-house legal resource for Sun-Diamond and the cooperatives for such matters as reviewing contracts and handling litigation. Sun-Diamond has a government relations and public affairs staff working with officials in Sacramento, Washington, Brussels and Asia to represent the interests of our growers and the cooperatives. We are keenly interested in regulatory matters, trade and public policy issues which impact agriculture and our business environment. For example, a few years ago we had an opportunity to help shape the Market Promotion Program, a program which provides funds for expanding the sale of agricultural products into export markets. We also work with the USDA on such programs as purchases for the school lunch program.

The government relations and public affairs department also produces the annual report for Sun-Diamond and the cooperatives each year, as well as a quarterly 48-page grower magazine. Individually the cooperatives could not commit the resources necessary for producing publications of comparable quality. Sun-Diamond also sponsors joint seminars for the boards of directors from the member cooperatives to educate them on current issues which affect our business, such as the environment, water policy and new food labeling requirements.

To better understand the functions which remain at the cooperative level, let us refer to Figure 9.3, the organization chart for Sunsweet. Sunsweet handles marketing, manufacturing, processing, packaging, financial matters, human resources and grower relations.

Sunsweet's major facility, almost a million square feet under one roof, is located in Yuba City, California. We have another plant in Fleetwood, Pennsylvania, a bottling facility for Sunsweet prune juice where we also co-pack juices for other companies. We also have arrangements with Ocean Spray to bottle Ocean Spray products in California. They in turn bottle prune juice for Sunsweet in several other parts of the country. Sunsweet can make acquisitions

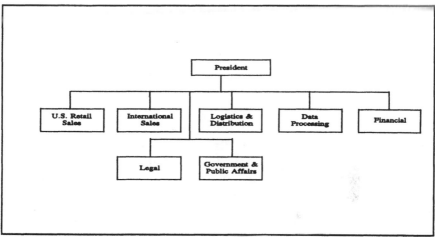

FIGURE 9.2 Sun-Diamond Organizational Structure

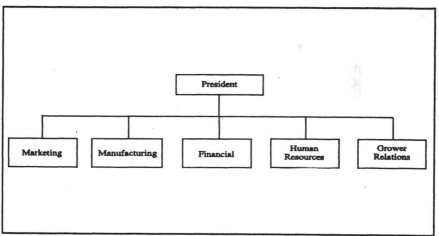

FIGURE 9.3 Sunsweet Organizational Structure

on its own. The facility in Fleetwood is one example. Each association has this capability, as does Sun-Diamond. However, we cannot compete against one another in the marketplace.

Sunsweet leases land to two companies at its Yuba City plant for cogeneration facilities. Fueled by natural gas, each generates electricity for 50,000 households annually in the northern part of the Sacramento Valley. The by-product is steam, and we use a million pounds of steam each day to process prunes. We entered into these agreements as Sunsweet, not as Sun-Diamond. We have that flexibility, as do the other associations.

At the member cooperative level we have separate political action committees for making political contributions. And Sunsweet has a subsidiary, Sunsweet International, a foreign sales corporation which provides certain tax advantages on export sales.

Now let's discuss joint sales representation, the primary reason for the cooperatives' acting in concert. Sun-Diamond provides joint representation for our consumer product line in the U.S. and internationally. The cooperatives' dried fruit products—raisins, prunes, figs, apples, peaches, apricots, mixed fruit—and nut products—walnuts, pecans, almonds, Brazil nuts, hazelnuts—fit together well into a full product line. Our brands have an outstanding merchandising presence in the dried fruit and nut category in supermarkets. They are found in the produce and baking sections.

A single agency for joint sales representation provides a more effective and more efficient sales arm than members would have standing alone. Through Sun-Diamond the cooperatives have a strong, professional sales organization with 20 sales professionals and 70 U.S. retail food brokers. Our size advantage allows us to select the best brokers available; our broker network is exceptionally strong. Because of the combined strength of our brands, we are the number-one or -two account for almost all of these brokers. They focus their attention on principals who earn them the most commission. If you have only one product to sell and you're dealing with busy people, your chance of getting their time is limited. If you can offer them a full range of products that have strong sales, you are going to get their time and attention. Getting their time and attention is the only way to ensure strong on-shelf merchandising as well as off-shelf displays to maximize your sales.

Although we go to the consumer at the brand level, we go to the trade as Sun-Diamond. This way we have greater clout. We offer the trade the advantage of one order, one invoice and one shipment for the full range of dried fruit and nut products. This is a real plus to the trade because it reduces the cost of doing business. We get additional shelf space and favorable positioning, which is the most important merchandising success factor. We are able to put together better promotional and merchandising programs. We get better displays off shelf and we attract cooperative ad support.

With this system we can also cover all classes of trade: grocery, mass merchandisers, and membership stores. A few years ago membership stores were small; today they are big business. Sun-Diamond's sales department projected their potential impact long before it happened. We were able to plan ahead so that now we have staff specifically designated to sell to each of these different classes of trade.

From time to time we've talked to other co-ops about joining us. I can assure you that any new partner would not be someone who produces cotton. The cotton people may fit with somebody else but they don't fit well with dried fruits and nuts. I can also tell you that any future partner should have a strong brand. We sell branded products, not generic commodities. The fit is all-important.

Geographic location has not been a problem for Sun-Diamond. Members have plants and offices in Fresno, Stockton, Yuba City and Pleasanton, California; Oregon; Illinois; Pennsylvania and Alabama. Our newest office is in Poland. We are opening an office there because their economies are beginning to emerge, and we are interested in new markets for our products. Geography makes no difference.

The last thing that I want to talk to you about are the critical success factors for partnership in a common marketing agency. Common marketing agency members need common goals and clear objectives for their partnership. There must be fair representation from the board of directors of each cooperative to govern the partnership. Partners must share similar marketing and sales strategies for product distribution, promotion programs to the trade and advertising. The partnership must provide consistent levels of service to each member without favoritism or discrimination. Member cooperatives are going to look to the common organization for reliability, for promptness, for dedication, for attention and for overall quality.

An alliance must be viewed as a long-term business project, but it must also be flexible. The time to form one is when the cooperatives are doing well. This provides an environment to consider proposals, resolve issues with an open mind, and to generally begin working together when pressures are at a low point. Because the business environment is dynamic, not static, provisions must be made to ensure that changes can be made quickly as needs dictate in the future.

For a common marketing agency to work requires compromise on many issues. It's not possible for members to maximize their position on each and every issue. But in many cases a member will sacrifice to serve the best interests of the overall partnership rather than stand alone. When the partners in such an alliance can realize benefits from joint activities while maintaining independence and flexibility, it can work exceedingly well.

I will conclude by emphasizing that the directors from each member cooperative must work together in a cooperative spirit to ensure the success of a partnership organization. The board of directors for each cooperative and the Sun-Diamond directors are personally committed to the long-term success of our alliance. A common marketing agency is no panacea, but it has worked for us. We believe we are all better off today than we were 12 years ago. We will continue working together, looking for ways to make it even better.

Notes

Mr. Giacolini appeared as a guest speaker at the NE-165 Workshop *New Strategic Directions for Agricultural Marketing Cooperatives*, Boston, Massachusetts, June 25, 1992. This paper is transcribed from that speech.

10

Cooperative Market Power and Antitrust with Application to California Information-Sharing Cooperatives

Richard J. Sexton and Terri A. Sexton

Agricultural cooperatives are often considered to be "countervailers" of market power. That is, the presence of a cooperative in a market may stimulate more competitive behavior from other market participants. This view of cooperatives has its roots in midwestern U.S. agriculture and is most often associated with Edwin Nourse (1922), although the framers of the Capper-Volstead Act of 1922 also embraced the concept. To the extent cooperatives perform this function, they create a social benefit.

About this same time a second school of cooperative thought originated in California through the work of Aaron Sapiro, who perceived a much more active role for cooperatives in coordinating market behavior than did Nourse. When one compares the agricultural economies of the midwest versus California, the logic and inspiration of both schools becomes apparent. Midwestern agriculture continues to be characterized by large numbers of relatively small farms producing standard grains, dairy, pork, or beef. Market conditions in these areas often have supported only one or a few local firms to sell farm supplies and buy and process farm products. Midwestern farmers were not in a position to control their markets, but they were vulnerable to monopoly/monopsony exploitation and so provided the inspiration for Nourse's school.

California agriculture is more diverse and involves more geographically concentrated production than the Midwest. California is the nation's leading producer of 65 crop and livestock commodities, often producing the lion's share of U.S. or even world production. California fruits and vegetables are often not storable, are usually produced by a relatively small number of farms, and are not subject to substantial geographic competition. Sapiro observed cooperatives operating in this structural environment and perceived the potential to exercise a degree of market control or market power when growers were united via a binding marketing contract in a single, commodity-oriented marketing association. Sapiro attempted to export this "California model" to other parts of the U.S. and to basic commodities such as grains and cotton but with little long-term success.[1]

A recent trend in California agriculture reflects the continued perception that cooperatives can be a device to achieve market power. Beginning in the early 1970s with the Central California Lettuce Producers Cooperative, growers of several fresh fruit and vegetable products have formed cooperatives that perform few, if any, actual marketing activities and do not take title to the product. Growers continue to market their own production, and the cooperative's only function is to share market information and establish pricing strategies. These cooperatives represent clear attempts to exercise market power through a cooperative. Whether they can be successful is yet unknown, as is the public policy question of whether they should be allowed to succeed assuming they have the potential to do so. What does seem clear is that these cooperatives will provide more impetus for attempts to limit cooperatives' exemptions from antitrust based on the Capper-Volstead Act.[2]

The purpose of this paper, therefore, is to analyze the potential exercise of market power by cooperatives with specific reference to California's new information-sharing cooperatives. The next section reviews briefly cooperatives' antitrust exemption under the Capper-Volstead Act. Next the economic theory of cartels is applied to conceptually analyze cooperatives' potential to exercise market power. Finally, the emerging phenomenon of information-sharing cooperatives is discussed with special emphasis on the lettuce producers' cooperative.

The Capper-Volstead Act

Beginning in the 1970s and continuing to today, cooperatives' exemptions from antitrust under the Capper-Volstead Act have been challenged by those who believe that certain cooperatives have abused their protection under Capper Volstead and have exercised market power to the detriment of consumers. In this section we briefly review the Capper-Volstead legislation and subsequent judicial interpretations.[3]

The Capper-Volstead Act consists of two sections. The first section authorizes the existence of agricultural marketing cooperatives that meet certain nominal restrictions (see, for example, Jesse *et al.* 1982). Section 1 was deemed necessary because in its absence the horizontal coordination of farmers in a cooperative and their associated activities, such as price setting and joint marketing, could be construed as a *per se* violation of the antitrust laws. Section

2 of the Act authorized the Secretary of Agriculture to investigate and order the cessation of monopolizing activity by a cooperative that "unduly" enhanced price. This provision has never been enforced.

The Act gives cooperatives the right to exist and compete as entities in the same fashion as ordinary corporations. Cooperative bargaining is also protected under the Act. Cooperatives may attain market power through voluntary association of farmers, but attempts to acquire a monopoly position through predatory practices and other anticompetitive activities have been judged outside protection of the Act and prosecutable under the Sherman Act. Similarly outside the scope of protection are joint ventures with noncooperative businesses. Mergers and joint ventures among cooperatives have never been challenged and are commonly believed to be protected under Capper-Volstead (Manchester 1982). However, critics argue that this protection was not intended and should not be provided (Federal Trade Commission Staff 1975, Baumer, Masson, and Masson, 1986). Similar controversy has emerged regarding the Secretary of Agriculture's failure to enforce Section 2's undue price enhancement clause. Baumer, Masson, and Masson (1986, p.251) argue that "at great cost to the consuming public, the Secretary of Agriculture has shirked his responsibility to supervise cooperative pricing," while others believe failure to initiate action merely reflects the reality that cooperatives lack the opportunity to exercise market power (Wills, 1985). The next section evaluates the bases for these conflicting claims.

Conceptual and Empirical Evidence on Cooperatives' Market Power

Conventional wisdom has been that cooperatives are ill-suited to attain market power. Senator Capper, during the debate on Capper-Volstead argued that a farmer's monopoly was impossible because, if price was set too high, the outcome would self-destruct due to over production in the ensuing years. Additional production in this context may occur from co-op members, nonmembers, new entrants, imports, or substitute products. Galbraith (1964) observed this same point and further argued that cooperatives were fatally flawed in the pursuit of market power due to (1) the voluntary nature of the association, (2) lack in almost all cases of complete inclusion of production in the cooperative, and (3) lack of absolute control over the decision to sell. Galbraith believed these factors led to the downfall of the Sapiro-style cooperatives in California. Torgerson (1978) has further noted that, because marketing cooperatives usually sell only a single product line, they lack the potential of for-profit conglomerates to pursue market power by cross subsidizing one product line with revenues from another.

Some antitrust lawyers and economists, however, have found this logic to be less than compelling. The Federal Trade Commission staff (1975) argued that cooperatives were dominant in the U.S. milk industry and several fruit, nut and vegetable industries: citrus, cranberries, Concord grape products, cling peaches, pears, walnuts, almonds, raisins, prunes, and winter celery. Except for the milk industry the FTC staff, however, presented no evidence that cooperative presence in these industries led to an exercise of market power.

The FTC staff argued for a strict and limited interpretation of cooperatives' antitrust immunity under Capper-Volstead. The National Commission for the Review of the Antitrust Laws in 1979 similarly recommended that mergers and joint arrangements among cooperatives be subject to the antitrust laws and urged greater attention to the undue price enhancement provision.

Using milk markets as a reference point, Baumer, Masson, and Masson (1986) have developed economic arguments to counter the logic that cooperatives are fundamentally incapable of exploiting market power. They note initially that demand for fluid milk is inelastic, and thus only a small percentage reduction in sales can generate a large percentage increase in price. Indeed demand for a number of agricultural products is similarly inelastic, but demand is only half of the output control quotient. If supply is *elastic* any increase in price will generate a large response in desired sales, complicating any supply control plans.

Supply control is also enhanced if there is an elastic-demand secondary market where excess production can be diverted (in other words, if a price discrimination scheme can be implemented). In dairy, the manufactured milk products market serves this function. Fruit and vegetable markets which have both fresh and processed product markets usually exhibit a similar pattern. Cooperatives can achieve nominal supply control by restricting membership and also possibly restricting the deliveries of members. Indeed failure to adopt either of these two policies is compelling evidence that a cooperative is not pursuing market power (Jesse *et al.*, 1982). The vast majority of U.S. cooperatives have open membership policies (Youde, 1978), and it is a cooperative tradition not to restrict member deliveries but, rather, to act as a "home" for members' production.

These prevailing cooperative policies have been cited by some commentators as evidence that cooperatives are not pursuing market power. This type of evidence is not persuasive, however, because in the industries where the FTC staff cited the presence of a dominant cooperative, both membership restriction and restriction of member deliveries are common practices.

In evaluating the potential success of these supply management activities, the key issue is whether the excess production will reach the market despite the cooperative's efforts to prevent it. In general, the answer would seem to be affirmative for the reasons that Galbraith cited. In no instance does a U.S. cooperative have complete hegemony over a market. Thus, nonmembers will generally be able to find alternative market outlets as will members for the portions of their production not handled by the cooperative. Moreover, those selling outside the cooperative are able to free ride on the cooperative's supply management activities and, thereby, earn a higher return than cooperative members.

Cooperatives and Marketing Orders

These considerations would often represent fatal flaws in cooperatives' plans to exercise market power if it were not for the Agricultural Marketing Agreement Act (AMAA) passed by Congress in 1937 which authorizes producer groups to operate marketing orders. These orders give the force of law to the supply management and market allocation programs that cooperatives are normally unable to achieve on their own due to free riding. Indeed, every analysis published to date suggesting market power abuses by cooperatives is actually an analysis of the anticompetitive effects of marketing orders, especially as they apply to milk marketing (FTC staff, Masson and Eisenstat 1980, and Baumer, Masson, and Masson, 1986). Cooperatives are implicated in the anticompetitive effects of marketing orders because they are often instrumental in promoting marketing orders, and the AMAA authorizes qualifying cooperatives to vote their members' votes as a bloc in marketing order referenda.[4]

Federal milk marketing orders regulate the flow of milk into fluid (class I) and manufacturing (class II) uses and define a number of geographic areas or pools for purposes of enforcing the regulations. Baumer, Masson, and Masson (1986) argue that cooperatives, through a variety of anticompetitive devices, have been able to extract price premiums above the federally mandated class I price by restricting product flows into the class I market.

To accomplish this end, cooperatives initially acquired large market shares in certain Federal market order areas through a wave of cooperative mergers in the 1960s. These mergers, if undertaken by ordinary corporations, could have been challenged under the U.S. Clayton Act but were allowed to proceed under the belief that cooperative mergers are protected under the Capper-Volstead Act.[5] To prevent free riders from undermining this market allocation scheme, cooperatives were accused of pushing exclusionary deals on buyers and enforcing long-term contracts with growers that effectively prevented either group from switching to an alternative handler. It was further suggested that predatory pricing was used to fight entry.

Finally, Baumer, Masson, and Masson (1986) argue that economies of scale created an entry barrier to prevent farmers from defecting from the existing cooperative structure. Long-term grower contracts with staggered expirations reinforced this effect by making it difficult for an entrant to capture transportation-cost economies by simultaneously signing up several growers along the same route.

In evaluating the various claims concerning cooperatives' and market power, a number of points are relevant. Cooperative mergers in the dairy industry, accomplished under the protection of Capper-Volstead, contributed to cooperatives' market share, but the abuses cited by cooperative critics could not have been accomplished without the aid of marketing orders. Second, the anticompetitive activity cited by Baumer, Masson, and Masson (1986) is not activity protected under Capper Volstead, and dairy cooperatives indeed were prosecuted for this activity under the antitrust laws (*Maryland and Virginia Milk Producers Assn v. U.S.* 362 U.S. 458, 1960).

Volume regulation through marketing orders implies a departure from competitive market conditions, and there may be good policy reasons to restrict this type of marketing order activity.[6] It is far less clear, however, that cooperatives' exercise of rights conferred upon them by marketing order legislation should be a basis to challenge favorable public policy towards cooperatives and, in particular, to repeal or emasculate the Capper-Volstead Act.

Cooperatives' Market Power Without Marketing Orders?

The more interesting and germane theoretical question for evaluating California's information-sharing cooperatives is whether cooperatives can exercise market power without government intervention through a marketing order. The question in essence is whether a cooperative can act as a cartel, and it can be analyzed via the economic theory of cartels. Jacquemin and Slade (1989) have recently surveyed this literature and list four prerequisites to achieving cartel power: (1) an agreement must be reached; (2) because participants have incentive to cheat on any agreement that raises price, cheating must be detected; (3) cheating, once detected, must be punished; and (4) outside entry must be deterred.

Reaching agreement among many independent sellers as to a marketing strategy may not be easy. Indeed reaching an agreement on market strategy is a primary function of the information-sharing cooperatives. However, neither these cooperatives nor cooperatives in general have been able to bring all relevant production within their membership. Thus, full agreement is seldom if ever achieved, and outsiders are able to free ride on any agreement among co-op members. Because outsiders do not abide by the restrictions contained in the agreement, they do better than the cooperating growers, and this fact provides a basis for members to defect from the cooperative.

Detecting cheating hinges upon observing unexpected patterns in sales or price. When there are many sellers as is typically the case in agriculture, the effects of cheating on individual firms' sales may be difficult to detect. Similarly agricultural prices are often highly volatile, so price decreases cannot be easily attributed to cheating. These characteristics of agricultural markets make detecting cheating diffioult and, hence, successful collusion less likely (Green and Porter, 1984).

Firms that cheat on cartel agreements make short-term gains. The key to punishing cheating and, thus, deterring it is to insure that long-term losses from cheating outweigh the short-term gains. A key feature of cooperative organizations is their legal right under Capper-Volstead to sign binding marketing agreements among members. Such agreements among ordinary corporations, of course, would be illegal.[7] These agreements need not be adhered to, but if the penalties for breach of the contract are stringent enough and the probability of detection is high enough, it is rational for the individual members to abide by the agreement. The ability to sign binding marketing agreements through cooperatives thus facilitates the exercise of market power. Indeed, binding marketing agreements were a cornerstone to Sapiro's school of cooperative thought (Cotterill, 1984, p. 40).

Preventing entry appears to be a compelling obstacle to cooperatives' use of Capper-Volstead to exercise market power. Because cooperation is voluntary, growers have incentive to remain outside the cooperative and free ride on its attempts to restrict volume. Members, in turn, have incentive to defect once their cooperative contracts expire. Further, barriers to *de novo* entry into production of the cartelized products is typically low in agriculture and may be accomplished in many cases simply by shifting acreage into the product or products in question. Tree fruit and nut crops with five to seven year maturities, however, provide a natural and relatively immutable short-run barrier to entry. The sunkenness of the investment in tree stock in these industries is also a barrier to entry.

As Baumer, Masson, and Masson (1986) argued, barriers to entry at downstream processing stages may frustrate attempts to enter at the farm production stage. However, the farm production sectors where producer numbers are low enough to make cartelization realistic are fresh fruits and vegetables which require relatively little processing and, hence, have modest economies of size in marketing.

Cooperatives like any other firm may undertake strategic actions such as exclusionary agreements or predatory pricing to deter entry. A considerable debate exists as to the effectiveness of these strategies, but in any event they are illegal under U.S. antitrust law and, as noted, Capper-Volstead does not immunize cooperatives from prosecution for these violations.

To summarize, the theoretical basis for market power abuses under marketing orders is strong, and cooperatives, particularly in the dairy industry have exploited the opportunities afforded them under marketing orders. These abuses are best dealt with through analysis of marketing order legislation rather than through legislation, such as Capper-Volstead, relating to cooperatives. The more intriguing theoretical proposition relative to policy towards cooperatives is whether the opportunity to sign legal, binding marketing agreements under

Capper-Volstead affords cooperatives the potential to exercise cartel power irrespective of the presence of a marketing order. For at least some fruit and vegetable commodities, production has become concentrated in the hands of sufficiently few producers to make this a realistic possibility.[8] Entry barriers into production of these crops is apparently low, although in some cases the geographic regions where the crops can be grown is limited, making the viability of cartel power an unsettled question from a theoretical perspective.

Empirical Evidence

Evidence suggesting market power abuses by dairy cooperatives operating under the auspices of marketing orders is strong. Prior to the initiation of antitrust proceedings against the leading dairy cooperatives, it was estimated that cooperatives raised retail milk prices from 7-10 cents per gallon with an annual social loss of $71 million (Masson and Eisenstat, 1980).

Sexton, Kling, and Carman (1991) found evidence suggesting that Florida celery growers succeeded in exercising market power in nearby terminal markets where they were insulated from California competition. Although Florida celery is marketed through a cooperative, this industry is also governed by a marketing order. We are aware of no evidence from industry studies that suggest cooperatives have enhanced price outside the presence of a marketing order.[9]

Two studies have measured the impact of cooperatives on market price. Wills (1985) analyzed prices for cooperative brands relative to prices for leading noncooperative brands. In all cases the prices for cooperative brands were lower than prices for comparable noncooperative brands with similar advertising and market shares. As Wills himself noted, however, criticisms of cooperative market power have not been levelled against cooperative branded products. In particular, it is commonly recognized that failure to establish strong brands is a general weakness of cooperative marketing, so this finding is not surprising.

In contrast to Wills' study which analyzed brands across industries, Haller (1992) studied the impact on prices of cooperatives' participation in a single market—cottage cheese—for 47 metropolitan areas. The results of Haller's analysis reveal that co-op brands are likely to be priced lower than non co-op brands, *ceteris paribus,* and also that a yardstick-of-competition effect apparently exists in the market. The presence of one or more cooperatives in the market was associated on average with a 4.7 cent (4.1%) decline in brand prices in the market, although an unknown portion of this effect is caused by the cooperative brands themselves being cheaper.

A related study by Petraglia and Rogers (1991) examines the effect of cooperative participation on relative price-cost margins in 136 U.S. food and tobacco industries. Cooperatives' participation in the markets was measured as the percentage of value added in the industry contributed by the 100 largest cooperatives in the U.S. The results show a negative and significant relationship between cooperatives' share and the price cost margin. A 10 percentage point increase in sales attributable to cooperatives was estimated to result in a two percent decrease in the relative price-cost margin.

Interpretation of this result is somewhat clouded because the price-cost margin may include both monopsony and monopoly power components. If cooperatives cause more competitive pricing in the procurement of raw agricultural products, relative margins would fall due to diminution of monopsony power (Nourse 1922, Sexton 1990). Also, if cooperatives were less efficient than for-profit corporations as some critics claim,[10] this effect would contribute to a negative correlation between the price-cost margin and cooperatives' shares. Therefore, it is not possible to separate the effect of cooperatives on output prices from their effect on monopsony power, or the relative efficiency of cooperatives.

California's Information-Sharing Cooperatives

Information-sharing cooperatives are a relatively new breed of cooperative that to date are confined exclusively to California. These cooperatives are organized under the usual statutes governing cooperatives, and the courts have determined that their activities are covered under the Capper-Volstead Act. However, these cooperatives perform no handling or other traditional marketing activities for their members' products, nor do they perform a collective bargaining function for their members. Rather, they exist as devices for their members to communicate, share information on production plans and other market information, and formulate pricing strategies. Simply put, these organizations perform many of the traditional functions of a cartel, although in practice they usually have not formed explicit pricing rules for their members.

This type of cooperative organization exists now in California for melon producers in the western San Joaquin and Imperial Valleys (the California Cantaloupe Growers Assn.), kiwifruit producers (the Kiwifruit Marketing Association), table grapes (the Coachella Grape Growers), fresh peaches, plums, and nectarines (the Associated Fruit Producers' Cooperative), mushrooms (the California Mushroom Growers Assn.), and in the lettuce industry the grandaddy of them all—the Central California Lettuce Producers Cooperative (Central). With the exception of Central, all of these associations have formed and began operations within the past four years. These organizations are receiving considerable publicity in the California farm press, and similar organizations in other fresh fruit and vegetable industries may spring up in the coming years.

Although this type of cooperative will probably not spread beyond the fresh fruit and vegetable industries, its activities are of general interest because they stretch the applicability of the Capper-Volstead Act to its limits. The rise to prominence of these cooperatives comes at a time, as noted, when some influential voices have been calling for a re-examination of Capper-Volstead. Although there seems to be no immediate threat to Capper-Volstead, the activities of these cooperatives are certain to provide fodder for those who wish to restrict the Act.   Thus, it is important to examine the activities of these cooperatives to evaluate their role in stimulating or stultifying competition and enhancing or diminishing the well being of farmers and consumers.

In undertaking this task, we focus on the lettuce cooperative, referred to as Central.   Because of Central's pivotal role in founding the California information-sharing cooperative movement, its sister cooperatives are often called "lettuce" cooperatives.   Because the other information-sharing cooperatives are still in their nascent stages, only Central provides a sufficient history to analyze the activities and performance of an information-sharing cooperative.

The Legal Disputes

Lettuce is grown extensively in central California's Salinas-Watsonville area in the summer and fall and in the Imperial Valley in southern California during the winter and early spring.   California is the dominant lettuce producer, depending upon the season, accounting for between 70 and 90 percent of the domestic supply.   Other producing regions of note include Arizona, Florida, New York, New Jersey, and Texas.   Imports from Canada and Mexico are minor.

Lettuce is a highly perishable crop which must be harvested and shipped within three to four days of ripening.   Lettuce is packed, usually unwrapped, into 24 head cartons and hauled by truck to a cooling facility, where it may be examined by buyers.   Lettuce from California is now shipped almost exclusively by truck, although rail shipments were common in the 1970s to early 1980s.

Most lettuce is sold at the shipping point on an FOB basis direct to wholesalers or major retailers.   Other lettuce, however, is consigned to wholesalers, who sell in major terminal markets (Atlanta, Boston, Chicago, New York, and Los Angeles, for example).   Consignment sales are on a profit-sharing basis.

In times of comparative oversupply shippers traditionally have made "distress" consignments when lettuce that could not be sold on a FOB basis was consigned to a wholesaler to sell at the best price possible.   Alternatively, shippers without a sale might "roll" eastbound carlots of lettuce and attempt to sell them enroute.   Another practice in times of oversupply has been to "no-bill" loads, that is, to hold an unsold car over for another day in an attempt to make a sale.

Lettuce plays an integral role in the American diet and, hence, has an inelastic demand.   This condition plus an unstable supply have caused highly volatile prices in the industry. Weekly price fluctuations of up to 800 percent have been experienced.

It was in this market environment that 22 central California lettuce grower-shippers formed Central in May 1972. The growers signed identical marketing contracts with the cooperative in June 1973. According to the agreement its purpose was

> preventing the demoralizing of markets resulting from dumping and predatory practices; mitigating the recognized evils of a marketing system under which prices are set for the entire industry by the weakest producer.

The agreement, which was limited to the Salinas-Watsonville (summer-fall) marketing season[11] imposed the following requirements on members:

1. reporting all relevant production information, including plantings, expected harvest dates, and volumes.
2. establishing prices within the limits of weekly or daily ceiling or floor prices established by the Cooperative.
3. agreeing to ship only on terms authorized by the Cooperative. In particular no open consignment sales or unsold rollers were allowed. So-called "price protection" was also prohibited.[12]
4. Reporting delinquent accounts and chronic complainers to the Cooperative.

Importantly, the members of Central continued to negotiate and make agreements with buyers independently; Central assumed no direct marketing or bargaining role. During the 1973 marketing season members of Central shipped about 20,200,000 cartons of lettuce, 64 percent of the total production from the Salinas-Watsonville marketing area, making it a considerable force in the lettuce industry.

The facts concerning Central's organization and operation are not in substantial dispute. It was and is a horizontal organization of lettuce producers intended to exercise market control, including price control, by limiting competition among its members. As such, unless Central's activities come under the protection of the Capper-Volstead Act, or Section 6 of the Clayton Act, they represent a clear violation of the Sherman Act's antimonopolizing provisions.

The issue of Central's qualification for Capper-Volstead protection was adjudicated in two separate legal proceedings that evolved through the mid 1970s. In June 1974 the Federal Trade Commission (FTC) issued a complaint against Central alleging violation of Section 5 of the FTC Act (essentially equivalent to violating Section 1 of the Sherman Act). Central suspended operations in November 1974 in response to the complaint. An administrative law judge (ALJ) entered an initial decision on March 13, 1975 sustaining the FTC's complaint and ordering the dissolution of Central. This decision was appealed to the full Commission, which dismissed the complaint in an order issued July 25, 1977 (Federal Trade Commission Decisions, Docket 8970, 90 F.T.C. 1977). Central resumed operations in May 1978, continuing to have about 20 members through the early 1980s.

During this same time period a private antitrust suit filed against Central by Northern California Supermarkets alleging Sherman Act Section 1 violations was proceeding through the Federal courts. This case was resolved in favor of Central by summary judgment of the District Court in January 1976 (*Northern California Supermarkets v. Central California Lettuce Producers Cooperative,* 413 F. Supp. 984, 1976). This decision was upheld by the appeals court (580 F. 2d 369, 1978), and certiorari was denied by the U.S. Supreme court (439 U.S. 1090, 1979).

Thus, except for the initial decision by the ALJ, all judicial proceedings concerning Central's antitrust exemption were resolved in its favor. In finding against Central, the ALJ ruled that Central performed none of the *collective* marketing practices described in the Capper-Volstead Act and, therefore, its activities constituted illegal price fixing. He read into the legislative intent of Capper-Volstead the desire to establish countervailing power by allowing farmers to aggregate to compete on more equal terms with oligopsony buyers. In his view Central's purpose was not to countervail market power but, rather, to promulgate it through, among other methods, price fixing. He found it untenable that Congress would have intended to sanction the creation of monopoly power for farmers.

As the Commission ruled upon appeal of this decision and as the Federal Courts also ruled in the *Northern Supermarkets* case, the ALJ's decision rested on rather untenable reasoning. Neither Capper-Volstead nor the Clayton Act, Sec. 6 specify a minimum threshold of collective activity that a cooperative must perform to qualify for exemption. The 9th Circuit Court had recently ruled that mere collective bargaining was sufficient activity to qualify for the exemption in *Treasure Valley Potato Bargaining Assn. v. Ore-Ida Foods, Inc.* (497 F.2d 203, 1974), and the Supreme Court had denied certiorari (419 U.S. 999, 1974). The ALJ found a meaningful difference between the collective cooperative bargaining undertaken in *Treasure Valley,* to which price setting was a necessary component, and the mere price control with independent marketing performed by the members of Central. However, this comparison was termed "a distinction without a difference" by the Court in *Northern Supermarkets.*

The most compelling logic supporting the District Court opinion and the Commission in overturning the ALJ's decision was the observation that the members of Central were clearly entitled to form under Capper-Volstead a much more tightly organized cooperative than in fact they had achieved under Central. That is, if they had formed a collective bargaining association or made sales as a single entity through a marketing cooperative, their collective activities would have indisputably been protected. Given that concerns over Central's opportunity to exercise market power stimulated the FTC's complaint, it would have been anomalous to stimulate, through the dissolution of Central, an even tighter and potentially more powerful producer organization.

The ALJ recognized this point but argued that the members of Central may not have been able to sustain a tighter organization. He suggested, moreover, that Congress ought to consider limiting the scope of Capper-Volstead, specifically to eliminate exemption for large agribusiness firms like the members of Central. This suggestion remains topical, and the issue of who constitutes (or who *should* constitute a farmer in terms of qualification for Capper-Volstead protection is viewed by many experts as a key issue to be resolved in the judicial and/or legislative arenas.

Analysis

Given the muddled state of affairs that would have emerged had Central been found in violation of the antitrust laws, the rulings by the FTC and the Federal Courts were correct and sensible. The situation that emerges as a consequence of these and other court interpretations of Capper-Volstead and the Clayton Act, Section 6 is that a cooperative organization of farmers can control the selling side of a market to the point of having an exclusive monopoly. Such power cannot be acquired, however, through anticompetitive acts such as predatory pricing and exclusive dealings (*Maryland and Virginia Milk Producers Assn v. U.S.*).

The question that remains is whether this state of affairs is desirable as a matter of public policy. Commentators ranging from the ALJ in the *Central* case to former FTC Chair Oliver and U.S. Senators Bradley and Metzenbaum have their doubts. The reason for concern is that agriculture has evolved dramatically since the time of Capper-Volstead when countless small farms produced agricultural products and were often at the mercy of middlemen. Undoubtedly the framers of Capper-Volstead did not envision the 1990s agriculture, where 15 growers in a single cooperative control Florida's entire celery crop or where 40 or fewer California growers can dominate production in some fresh fruits and vegetables. The movement of growers in several commodities to form "lettuce" cooperatives in recent years compels examination of the prospects for and desirability of farmers having collective monopolies in the sale of these products.

We began to analyze the potential for co-op market power in the prior section with our investigation of the prerequisites to cartel power. Any cartel arrangement that succeeds in raising price above competitive levels is vulnerable to member defection, so the primary cartel-power advantage accruing to a Capper-Volstead cooperative is the opportunity to sign legally binding agreements which contain specific penalties for violation of the agreements. The alternative is to enforce cooperative behavior by credibly threatening a regime of noncooperative pricing which is harmful to both the deviating firm and those executing the punishment strategy (Green and Porter 1984). Thus, Capper-Volstead lowers the cost of punishing violations of a cooperative agreement among members.

If an agreement can be reached and cheating can be detected, then deterring entry remains as the primary impediment to the exercise of cartel power by farmers through cooperatives.[13] In general, scarcity of appropriate land upon which to grow the commodity is the only potential entry barrier that comes to mind for annual crops like lettuce.[14] Thus, except in cases where land represents a limiting resource, farmer monopolies are not a policy concern in these industries because attempts to raise price above competitive levels will induce entry.

Can farmer cartels represent good economic policy? Three arguments can be made on their behalf. The first is the *countervailing power* argument. This concept is implicit in the legislative intent of Capper-Volstead, although the theory is most commonly associated with Galbraith. McCormick (1991) has argued recently that lettuce cooperatives are indeed an appropriate attempt to countervail increasing buying power in the produce industry. This concentration is a consequence of mergers in the food industry and joint purchasing arrangements among buyers. McCormick indicates, for example, that there are only 10 buyers for the fresh fruits sold by AFPC.

In contrast to the anticompetitive effects from exercising cartel power, the exercise of countervailing power portends favorable economic consequences through the diminution of buyers' monopsony power. Thus, even if it is concluded that a cooperative has succeeded in raising price, it must also be established that the price was increased above rather than to the competitive level before concern is justified.

The second possible social benefit from a cooperative having price control is the moderation of price fluctuations. This objective was clearly part of Central's initial operating plan, and it has been cited as a primary effect of Central's operation by Garoyan, Kinny, Pisani, and Skinner (1983) in the only attempt to date to statistically analyze Central's impact on the lettuce market.

It is unclear based on the Garoyan *et al.* analysis how effective Central has been in stabilizing prices. Although the cooperative has established price ceilings and/or floors, presumably these must be accompanied by volume controls to have much impact, and volume control *post planting* is extremely difficult for this highly perishable commodity. Members of Central exchange planting and harvest information which may help them to plan a better flow of product to the market.

Questions persist as to the price stabilizing role, however.

1. Can a "lettuce" cooperative organized along the lines of Central succeed in stabilizing price?

2. Is price stabilization (rather than profit maximization) likely to be an important goal for cooperatives like Central?
3. Under what conditions do producers and/or consumers benefit from price stabilization in the fresh produce industries?

The final benefit imputed to cartels concerns industries with high fixed costs and, hence, a large minimum efficient operating scale relative to the total demand in the industry. These industries can support only a few firms, and no competitive equilibrium may exist. Specifically, all firms will likely be unable to operate at the minimum points on their unit cost curves. In other words, market demand at the value of minimum average cost is unlikely to be an integer multiple of the efficient output level. However, any firm not operating at the efficient scale can be undercut by a rival, and so there is no equilibrium.

Several authors including Telser (1978) and Bittlingmayer (1982) have argued in these cases that firms may collude and reach an agreement to set prices above marginal cost as a way of achieving a stable industry outcome. Collusion and modest price enhancement, these authors argue, is preferable in these industries to alternative solutions, such as mergers, to the instability caused by nonexistence of the competitive equilibrium. This theory does not seem to apply to the produce industries where fixed costs are not high and the industries to this day continue to support a larger number of firms than in the industries discussed by Telser and Bittlingmayer.

The main conclusion to draw from this analysis is that arguments in favor of grower cartel power and price control in the produce industries rest heavily on the countervailing power argument. Indeed this may be a powerful argument, given the level of buyer power now observed in many of these industries.

Conclusions

This paper has conducted a conceptual analysis of the issue of cooperatives' market power exercised under the protection of Section 6 of the Clayton Act and the Capper-Volstead Act. The analysis was prompted by the recent organization of several cooperatives in California which apparently have the exercise of market control or cartel power as their primary objective. Analyses of cooperatives' market power to date have focused on cooperatives' role in the operation of marketing orders and on the dairy industry. Here any market power abuses seem tied directly to the provisions of the marketing orders and have little, if anything, to do with the protections afforded by the Clayton Act and the Capper-Volstead Act.

In studying the question of whether agricultural cooperatives are suited to exercise market power without the accompaniment of a marketing order, our analysis pinpointed cooperatives' ability to sign legally binding agreements concerning prices and production as their main device to surmount the problems facing cartel organization. The key impediment to successful exercise of cartel power through a cooperative would appear to be the prevention of entry, because entry barriers into agriculture are typically low. Incentives of members to defect from successful agreements also present a potential problem.

Whether the exercise of market control through a cooperative can have socially desirable consequences was concluded to depend primarily upon whether the cooperative was countervailing the monopsony power of buying firms—raising price towards the competitive level—or exercising monopoly power in its own right by enhancing price above the competitive level. No empirical evidence has been produced to date to suggest that cooperatives have successfully exercised market power in industries without volume regulation through a marketing order. Our short-term goal in this continuing research is to empirically analyze pricing behavior in the lettuce industry to determine the lettuce cooperative Central's influence on the level and stability of prices in the industry.

With the upsurge of the California information-sharing cooperatives, the issue of cooperatives' market power must be addressed. It is not clear that the present state of affairs—absolute protection of price fixing under Capper-Volstead is desirable. One possible modification to consider is to replace this *per se* protection with a *rule of reason* criterion whereby farmers' right to organize into cooperatives remains immutable, but pricing decisions are subject to scrutiny for that small subset of cooperatives that have the structural potential to exercise market power. Here the key criterion in evaluating pricing decisions would be whether the practices were designed to countervail or promulgate market power. In essence, this type of evaluation is what section 2 of Capper-Volstead was intended to provide, but similar scrutiny could be provided within the general framework of the antitrust laws.

We are not sanguine, however, that attempts to limit the scope of Capper Volstead by refining or limiting the definition of who qualifies as a "farmer" under the Act will be helpful. Rather, attemps to excluse the large, vertically integrated grower-shippers who are members of Central could induce reconsition, such as vertical divestitures, of otherwise efficient organizational forms.

Notes

1. See Knapp (1973) for a historical account of the evolution of the Nourse- and Sapiro-style cooperatives and a description of their underlying economic and social philosophies. See Cotterill (1984) for an interesting attempt to synthesize the two schools.

2. For example, U.S. senators Bradley and Metzenbaum in August 1989 requested the General Accounting Office to study the Capper-Volstead Act, and Daniel Oliver, at the time he was chair of the Federal Trade Commission, called for elimination of cooperatives' exemptions under Capper Volstead (American Institute of Food Distribution 1988).

3. The Clayton Act, section 6 also protects farmers' right to organize nonstock cooperatives. Subsequent legal interpretations suggest that Capper-Volstead provides a broader range of protections than Clayton, so it is the focus of discussion here.

4. Federal marketing orders are proposed by producer groups to the Secretary of Agriculture, who is charged to investigate the need for an order. If this finding is affirmative, the order must be ratified by a vote of producers before it becomes effective. In cases where cooperatives control a majority of production and can cast their members' votes as a bloc, the cooperative effectively controls the marketing order.

5. The largest U.S. dairy cooperative, Associated Milk Producers, Inc. (AMPI), was formed by a series of 10 mergers, beginning in 1967 and grew through 54 additional mergers in the next several years. Masson and Eisenstat report that AMPI had over a 90% market share in 10 southern markets in 1973.

6. French (1982) surveys the potential pros and cons of marketing orders.

7. Collusion in these cases must be supported by noncooperative equilibrium strategies. See Jacquemin and Slade for a summary of this literature.

8. For example, about 15 growers dominate the Florida celery market. From 20-40 growers control much of the supply for many California fresh vegetables.

9. However, because little empirical research has been conducted on this topic, there is also not contrary evidence, i.e., that cooperatives have not raised price.

10. Studies suggesting cooperatives' departures from efficiency include Porter and Scully (1987), Hollas and Stansell (1988), Sexton, Wilson, and Wann (1989), and Ferrier and Porter (1991). Studies suggesting that cooperatives may outperform their for-profit counterparts include Babb and Boynton (1981), Lerman and Parliament (1990), and Parliament, Lerman, and Fulton (1990).

11. Agreements with Central now also extend to the Imperial Valley (winter-spring) marketing season.

12. Price protection is a practice in the produce industry whereby the shipper agrees to compensate the buyer if prices fall below the agreed upon FOB price while the shipment is enroute to its destination.

13. As discussed in the prior section, meeting these two requirements is a nontrivial issue. Cooperatives have almost never reached full agreement (100 percent membership) in an industry. Outsiders, of course, free ride on any agreement and contribute to its demise.

14. Strategic entry barriers erected by a cooperative such as predatory pricing are not protected actions under Capper-Volstead or the Clayton Act, section 6 (*Maryland and Virginia Milk Producers Assn. v. U.S.*).

References

American Institute of Food Distribution. 1988. *The Food Institute Report*, Dec., p. 10.
Babb, E.M. and R.D. Boynton. 1981. Comparative Performance of Cooperative and Private Cheese Plants in Wisconsin. *North Central Journal of Agricultural Economics*

Economic and Legal Analysis of the Antitrust Exemption for Agriculture. *Villanova Law Review* 31:183-252.

Bittlingmayer, G. 1982. Decreasing Average Cost and Competition: A New Look at the Addyston Pipe Case. *Journal of Law and Economics* 25:201-29.

Cotterill, R.W. 1984. The Competitive Yardstick School of Cooperative Thought. *American Cooperation 1984*, Washington, DC: American Institute of Cooperation.

Federal Trade Commission Staff. 1975. A Report on Agricultural Cooperatives. Federal Trade Commission.

Ferrier, G.D. and P.K. Porter. 1991. The Productive Efficiency of US Milk Processing Co-operatives. *Journal of Agricultural Economics* 42:161-173.

French, B.C. 1982. Fruit and Vegetable Marketing Orders: A Critique of Issues and State of Analysis. *American Journal of Agricultural Economics* 64:916-923.

Galbraith, J.K. 1964. *American Capitalism*, London: Hamish Hamilton.

Garoyan, L., W. Kinney, J. Pisani, and R. Skinner. 1983. The Value and Use of Economic Information for Cooperative Marketing. unpublished manuscript, Department of Agricultural Economics, University of California, Davis.

Green, E. and R.H. Porter. 1984. Noncooperative Collusion under Imperfect Price Information. *Econometrica* 52:87-100.

Haller, L.E. 1992. Branded Product Marketing Strategies in the Cottage Cheese Market: Cooperative Versus Proprietary Firms. In *Competitive Strategy Analysis in the Food System*, ed. R.W. Cotterill, Boulder, CO: Westview Press.

Hollas, D.R. and S.R. Stansell. 1988. An Examination of the Effect of Ownership Form on Price Efficiency: Proprietary, Cooperative and Municipal Electric Utilities. *Southern Economic Journal* 55:336-350.

Jacquemin, A. and M.E. Slade. 1989. Cartels, Collusion, and Horizontal Merger. In *Handbook of Industrial Organization*, ed. R. Schmalensee and R. Willig, Amsterdam: North Holland.

Jesse, E.V., A. Johnson, B. Marion, and A.C. Manchester. 1982. Interpreting and Enforcing Section 2 of the Capper-Volstead Act. *American Journal of Agricultural Economics* 64:431-443.

Knapp, J.G. 1973. *The Advance of American Cooperative Enterprise: 1920-1945*, Danville, IL: The Interstate Printers and Publishers.

Lerman, Z. and C. Parliament. 1990. Comparative Performance of Cooperatives and Investor-Owned Firms in US Food Industries. *Agribusiness* 6:527-540.

Manchester, A. 1982. The Status of Marketing Cooperatives Under Antitrust Law, ERS-673, U.S. Department of Agriculture, Economic Research Service.

Masson, R.T. and P.M. Eisenstat. 1980. Welfare Impacts of Milk Orders and the Antitrust Immunities for Cooperatives. *American Journal of Agricultural Economics* 62:270-278.

McCormick, L.W. 1991. Is There a New Kind of Agricultural Marketing Cooperative on the Horizon? paper presented at the National Bargaining Conference, Monterey, California.

National Commission for the Review of Antitrust Laws and Procedures. 1979. Report to the President and Attorney General.

Nourse, E.G. 1922. Economic Philosophy of Cooperation. *American Economic Review* 12:577-597.

Parliament, C., Z. Lerman, and J. Fulton. 1990. Performance of Cooperatives and Investor-Owned Firms in the Dairy Industry. *Journal of Agricultural Cooperation* 5:1-16.

Petraglia, L.M. and R.T. Rogers. 1991. The Impact of Agricultural Marketing Cooperatives on Market Performance in U.S. Food Manufacturing Industries for 1982. Food Marketing Policy Center Research Report No. 12, University of Connecticut.

Porter, P.K. and G.W. Scully. 1987. Economic Efficiency in Cooperatives. *Journal of Law and Economics* 30:489-512.

Sexton, R.J. 1990. Imperfect Competition in Agricultural Markets and the Role of Cooperatives: A Spatial Analysis. *American Journal of Agricultural Economics* 72:709-720.

Sexton, R.J., B.M. Wilson, and J.J. Wann. 1989. Some Tests of the Economic Theory of Cooperatives: Methodology and Application to Cotton Ginning. *Western Journal of Agricultural Economics* 14:55-66.

Sexton, R.J., C.L. Kling, and H.F. Carman. 1991. Market Integration, Efficiency of Arbitrage, and Imperfect Competition: Methodology and Application to U.S. Celery. *American Journal of Agricultural Economics* 73:568-580.

Telser, L. 1978. *Economic Theory and The Core*, Chicago: University of Chicago Press.

Torgerson, R. 1978. An Overall Assessment of Cooperative Market Power. In *Agricultural Cooperatives and the Public Interest*, ed. B. Marion, North Central Regional Research Publication 256.

Wills, R. 1985. Evaluating Price Enhancement by Processing Cooperatives. *American Journal of Agricultural Economics* 67:183-192.

Youde, J. 1978. Cooperative Membership Policies. In *Agricultural Cooperatives and the Public Interest*, ed. B. Marion, North Central Regional Research Publication 256.

About the Book and Editor

As debates over NAFTA and GATT continue, agricultural trade issues remain at the forefront of policy and trade concerns. This timely study explores the evolution of agricultural marketing cooperatives within the framework of competitive strategy analysis.

Bringing together the diverse perspectives of academics and practitioners, this volume offers fresh insights into topics such as horizontal and vertical integration, product differentiation, strategic planning from a business perspective, and common marketing agencies as well as suggesting new strategic directions that cooperatives might pursue. This book will be useful for scholars of agricultural trade, economics, and business as well as agricultural marketing economists, policymakers, and industry practitioners.

Ronald W. Cotterill is director of the Food Marketing Policy Center and professor of agricultural economics at the University of Connecticut. He is the editor of *Competitive Strategy Analysis in the Food System* (Westview 1992) and author of numerous publications on the economics and technology of the food industry.

About the Contributors

Richard E. Bell is President and C.E.O., Riceland Foods, Inc.

Sanjib Bhuyan is a research assistant in the department of Agricultural and Resource Economics at the University of Connecticut, Storrs.

Julie A. Caswell is an associate professor of Agricultural Economics at the University of Massachusetts, Amherst.

Ronald W. Cotterill is a professor of Agricultural and Resource Economics and director of the Food Marketing Policy Center at the University of Connecticut, Storrs.

Robert A. Cropp is a professor and theory marketing and policy specialist in the department of Agricultural Economics at the University of Wisconsin, Madison.

William D. Dobson is a distinguished professor in the department of Agricultural Economics at the University of Wisconsin, Madison.

Earl Giacolini is Chairman of the Board, Sun-Diamond Growers of California.

Lawrence E. Haller is a research assistant at the Food Marketing Policy Center, Agricultural and Resource Economics, University of Connecticut, Storrs.

Bruce J. Reynolds is program leader for Strategic Management and Planning, Agricultural Cooperative Service, United States Department of Agriculture.

Tanya Roberts is an agricultural economist with the Economic Research Service, United States Department of Agriculture.

Richard T. Rogers is an associate professor of Resource Economics at the University of Massachusetts, Amherst.

Jeffrey S. Royer is an associate professor of Agricultural Economics at the University of Nebraska, Lincoln.

Richard J. Sexton is an associate professor of Agricultural Economics at the University of California, Davis.

Terri A. Sexton is a professor of Economics at California State University, Sacramento.

Randall E. Westgren is an associate professor and department chair of Agricultural Economics at McGill University, St-Anne-de-Bellevue, Québec, Canada.

Printed and bound by CPI Group (UK) Ltd, Croydon, CR0 4YY
23/10/2024
01778241-0009